CHORDATE EVOLUTION

CHORDATE EVOLUTION

M. A. Subramanian

MJP Publishers

Chennai New Delhi Tirunelveli

ISBN 978-81-8094-184-9

MJP Publishers

New No. 5, Muthu Kalathy Street,
Triplicane, Chennai 600 005

MJP 152 MJP's imprint: 2024

PREFACE

This book has been put forth as a general text book with the aim to provide a comprehensive and integrated account on chordate evolution. Chordata is taxon which includes the well known vertebrates such as fishes, amphibians, reptiles, birds and mammals including human beings and are well represented in marine, fresh water, and terrestrial habitats. The events of chordate evolution through geological time have been interpreted in this book based on the fossil record and available data. It is not an easy task to follow the evolutionary trends of chordates because different groups of vertebrate animals evolved simultaneously in separate lines. However, the facts of major evolutionary changes among the chordates and the related theories have been highlighted appropriately in the text. The characteristics of animals have been exemplified by a number of drawings selected from various sources of literature. The subject matter of the book has been written in an easily accessible language and style. No doubt that this manual will be highly valuable to the students and professionals in evolutionary biology and palaeontology in addition to any interested personalities.

I am much pleased to express my thanks to my family members and friends for their cooperation in the preparation of the manuscript. My heart felt thanks are due to MJP Publishers, Chennai for their encouragement and interest in publishing the book. Constructive criticisms and suggestions for the improvement of the text are immensely appreciated.

CONTENTS

1

INTRODUCTION

Chordates comprise a large group of animals possessing a notochord or a backbone throughout their life or at some stage in their development (vertebrates). They were established and diversified during early Palaeozoic times. Major groups of fishes appeared during middle and late Devonian periods. The first land vertebrates, namely the amphibians evolved from certain advanced fishes during the transitional period from Devonian to Mississippian. The reptiles evolved from the amphibians during Pennsylvanian period. Then two major vertebrate groups namely the birds and mammals appeared during Jurassic and Triassic times respectively from the reptilian group. In the process of continued evolutionary process, the intellectual mammal namely man evolved.

The study of chordate evolution is always impressive and fascinating as more current ideas have been laid down amidst the old hypotheses and fossil proofs. In general, evolution is the change in the characteristics of a population with a resultant splitting into small groups. Such groups would evolve independently and diversify ultimately into new species. The great diversity of world's life has always attracted man and so he has made many attempts to understand the processes and causes of this diversity. The earliest concepts of biologists on evolution are often inconsistent and there is no universal conclusions with regard to evolutionary events. In 19th century, evolutionists succeeded in demonstrating the occurence of evolution. While the phylogenetic approach was prevalent for the evolutionary aspects during the second half of 19th century, features of genetics and related biological disciplines allured many people in the 20th century.

The evolutionary trends of chordate animals and the causes of these evolutionary events have been well documentated by the biologists who have explained various facts in the form of theories. Currently, genetics and molecular biology are the major fields in the evolutionary biology so that the modern evolutionary concept revolves around natural selection and inheritance. Now the evolutionary biology especially the chordate evolution is considered to be an area of science with many practical applications. In general, the evolutionary concepts remain as the major arena in ecology, genetics, physiology and biochemistry including psychology, medicine and philosophy.

2

FOSSILS IN THE STUDY OF CHORDATE EVOLUTION

The fossils are the preserved remains of the animals and plants of the past that have become extinct. In general, they are petrified hard parts of animals which have transformed into stone. They may also include any hard parts that are preserved intact without petrification. The soft parts of organisms may also be fossilized occasionally. However, the complete preservation of extinct animals without alterations is rare.

The study of fossils is comparitively a new branch of science and the actual understanding of fossils has been fulfilled by Leonardo da Vinci and the interpretation of fossil record with reference to modern evolutionary concept starts with the work of Charles Darwin. At present, a large quantum of fossils have been discovered, studied and translated by palaeontologists all over the world. We can say that the knowledge of evolutionary history of life, especially of vertebrates, is now completely based on the fossil record. The collection and preparation of fossils to carry out the study is a laborious task. This is because the fossils are to be collected by special techniques, brought to the laboratory and made ready for investigation. Sometimes, large extinct animals have to be reconstructed with bones, soft parts such as muscles, nerves, and blood vessels including outer coverings (skin or scales) in relation to the knowledge of modern animals. In addition, many new forms are still to be discovered in order to fill many gaps in the evolutionary history of animals and plants. Any how the general aspects of vertebrate evolution has been well knowledged through the fossil record.

Kinds of Fossils

Fossils may be in the form of preserved bones or shells of animals or moulds /imprints in the rocks left by the animals/plants or foot prints of animals or preserved built up structures by the animals (nests or tubes). The most important kind os fossils include the petrified hard parts of animals and plants in rocks due to the burial of organisms in the sediment which is deposited layer after layer. Therefore, the oldest fossils occur in deepest layers and the recent ones are found in superficial layers. Sometimes, complete animals may be preserved as fossils very rarely as in the case of insects preserved in amber, mammoths frozen in ice, woolly rhinoceros in asphalt and so on. Fossils may also occur in the form of moulds and casts formed by the leaves and small marine animals. They, in general, include impressions of foot or leaf prints.

Role of Fossils in the Construction of Geological Time Scale

The earth's crust containing fossils is nothing but a register of geological time and its different layers represent the leaves of register in which the evolutionary history of organisms is found in the form of fossils. The fossil plants and animals of lowest rocky layers must have existed at an early period than those that are found at top layers. The rock which contain identical fossils would possess the same age. If the rocks contain simple plant and animal fossils, then they are older ones. But the recently formed rocks contain fossils of complex forms. These fossils play a vital role to understand the evolutionary trends of the ancestral history of animal groups. The fossils which occur in different strata differ among themselves but the fossils of adjacent layers are more alike. The more recent fossils have greater resemblances to the existing groups than those of the lower strata. The sediments containing the same types of fossils might have been formed at the same time and under similar conditions.

The oldest known rocks are the Isua rocks from south western greenland formed around 3700 million years ago but they preserve no fossils. The oldest to date known fossils are from the Apexchert rocks of north western Australia formed around 3465 million years ago. These fossils are found to include the Cyanobacteria. The possible Eukaryotes might have appeared 2000 million years ago. The Proterozoic fossil record documents most of the Eukaryote kingdoms including green plants. The multicellular metazoans seem to appear in the fossil record at the end of the Permian period.

Role of Fossils in Understanding the Process of Evolution

The fossils are highly valuable to understand the process of evolution in addition to the construction of the geological time scale. The fossils are considered as the 'written documents of evolution' contained in the rocks. It is found that the fossils of older rocks are more simple than those contained in new rocks and they differ from the living forms. On the other hand, the fossils of younger rock resemble the present day organisms more closely. This is suggestive of the fact that more advanced and complex organisms might have evolved from simple and more primitive ancestors. It is possible to have a knowledge with regard to the extinct plants and animals that lived in the past. For example, the dinosaurs, ammonoites, Trilobites etc., can be known only from the fossil record. The fossils help to determine the relationship among different groups of plants and animals. Some fossils form connecting link between two groups of organisms as in *Archaeopteryx* which serves as a connecting link between reptiles and birds. A complete sequence of ancestral forms of certain animal groups are let known by the fossil record as noticed in the pedigree of horse, elephant and camel. Above all, fossils clearly indicate the ancient geography as well as the wide fluctuations in the climatic conditions of the earth history.

QUESTIONS

1. What are fossils? Write an account on the importance of fossils in the construction of geological time scale.
2. How could fossils be useful in the process of evolution?
3. Write notes on the following:
   a) Fossil types
   b) Fossils and geological time scale.
   c) Fossils and evolutionary process.

3

GEOLOGICAL TIME

Our planet earth and its living organisms possess a long history. Many attempts have been made to estimate the age of earth by exploring the layers of rocks. The entire life span of earth is known as the GEOLOGICAL TIME during the course of which organisms exhibit a sequence of evolutionary events. The evolutionary changes of animals could be well known by the study of fossils that are found in different strata of rocks. With the knowledge of rocks and fossils, the geological time has been divided into definite intervals, each with significant events and different durations. The larger divisions of geological time are called as eras which are in turn divided into periods. Again, the periods are subdivided into lesser divisions called epochs. The arrangement of these divisions on the time scale in the order of their age is called 'geological time scale'. It has been constructed based on the correlations of the deposits all over the world via the study of fossils. The widespread geological disturbances evidenced in each period of time are called revolutions which determined the distribution of land and aquatic organisms and caused the decline of many of the groups.

The study of origin and evolution of animals necessiates a knowledge of earth's history or geological times. This can be achieved by knowing or estimating the age of rocks or geological strata in which the remnants of organisms of the past (fossils) are present. The major eras of earth's history are denoted as Archaeozoic era, Proterozoic era, Palaeozoic era, Mesozoic era and Cenozoic era.

Among them, the first two eras are found to exist prior to the beginning of the fossil record. As it is very difficult to interpret these times, this portion of the geologic time is referred to as the Precambrian times with a long span of time (more than 4500 million years ago). The Archaeozoic era is the 'period of primal life' characterized by having a time span of about 1000 million years with abundant graphite iron ore and lime stone deposits. During this time, a complete fossil record of organisms was not made available. Probably the primordial heterotrophic organisms might have evolved during this time. The Proterozoic era is the 'period of primitive life' with a time span of about 1500 million years. During this era, algae, marine protozoans, molluscs, annelids and other marine invertebrates were found as fossils. But most of these fossils were lost due to changes on the earth's crust. Because of these reasons, the Archaeozoic and Proterozoic eras have been collectively called as Cryptozoic era or Precambrian era. Based on the fossil record, the three major eras of earth's history in the order of their age are Palaeozoic era, Mesozoic era and Cenozoic era.

PALAEOZOIC ERA

The Palaeozoic era was the time of ancient life about 505 million years ago and lasted for about 300 million years. It has been divided into six periods namely Cambrian, Ordovician, Silurian, Devonian, Mississippian, Pennsylvanian and Permian Periods based on the order of age. Some Palaeontologists combine the Mississipian and Pennsylvanian periods into a single division namely the Carboniferous period. The Cambrian is the oldest period of Palaeozoic era (500–600 million years ago) and lasted for about 80 million years. The rocks of this period contained a large number of marine invertebrates especially Trilobites and branchiopods. Hence this period is referred to as the 'age of Invertebrates'. The Ordovician period (425–500 million years ago) lasted for about 65 million years and was characterized by the appearance of the first fishes, most probably the fresh water fishes. This period also evidences the presence of a number of invertebrate fossils such as corals, eurypterids, trilobites, molluscs, echinoderms, etc. The time span of the Silurian period was about

35 million years (405–425 million years ago). The terrestrial plants, wingless insects, marine arachnids and the first fishes appeared during this period. The Devonian period is considered as 'the age of fishes' as many fish species were abundant and dominated during this period (345–405 million years ago). This period lasted for about 45 million years and was characteristic for the appearance of the first amphibians. The Carboniferous period extended for about 50 million years (230–320 million years ago) and was evidenced for the origin of amphibians and winged insects, the appearance of stem reptiles, the dominance of marine echinoderms especially sea lilies and the spread of ancient sharks and amphibians. Permian is the youngest period in Palaeozoic era and lasted for 25 million years about 230–280 million years ago. It was characterized by the appearance of mammal-like reptiles and the extinction of several lines of amphibians.

Chordate Fauna of Palaeozoic Era

Devonian period is one of the most important times in vertebrate evolution and can be considered as a time of 'evolutionary explosion' that have occured on earth history. This is because, the jawless vertebrates (Placoderms and Acanthodians), sharks and bony fishes became well represented by the late Silurian and Devonian periods. Moreover, the amphibians became evolved at the end of Devonian. For the fishes, the Devonian period was a time of evolutionary advancement, as they were the dominant aquatic animals throughout the world. Thus the Palaeozoic era was a period of the first flowering of the vertebrates. Though the ostracoderms dominated and continued their evolution and the jawed vertebrates became well established, they faced competition from the early acanthodians and placoderms. The ostracoderms which flourished in early Devonian were reduced much and represented only as remnants in the transition from early to middle and late Devonian times. The acanthodians also became reduced whereas the arthrodires (placoderms) expanded and invaded the estuaries and seas. They were at the culmination of their evolution in late Devonian as the giant predators.

During middle Devonian period, the bony fishes (Osteichthyes) appeared and this event might be the reason for the decline and disappearance of the primitive vertebrates like ostracoderms and placoderms. The bony fishes evolved and diversified greatly during middle Devonian as fresh water species at the beginning and then as marine fishes. At this time, the group Osteichthyes differentiated into two lines namely dipnoans (lung fishes) and crossopterygians. As revealed by fossil record, the sharks appeared during upper Devonian times. At the end of Devonian period, the climax of vertebrate evolution is evidenced by the appearance of the first amphibians, the beginning of tetrapod history. The amphibians were dominant on the land during Mississipian period without any competition. The reptiles arose in Permian times so that the amphibians and reptiles lived side by side for a long time. But the reptiles were more efficient and better adapted than the amphibians. The large labyrinthodont amphibians disappeared completely from the land after the beginning of the evolution of dinosaurs. Since these reptiles became highly advanced, the proportion of amphibians decreased through Carboniferous and Permian periods. At this time, the primitive Cotylosaurian reptiles had completed their evolutionary development and were at the verge of extinction. Now the Pelycosaurs were succeeded by therapsids and mammal-like reptiles were in full expansion at the end of the Permian period. Thus starting from the Cambrian through Devonian to Permian periods, highly varied groups of fishes were found to inhabit the streams and the ponds of the continents and oceans and a number of diverse groups of amphibians in the watery environments of the land with the domination of most advanced terrestrial reptiles on land were present at the close of the Palaeozoic era.

Environmental Conditions of Palaeozoic Era

The primitive pre-Devonian vertebrates were found in continental fresh waters, rivers, streams and lakes so that their habitats would be the shallow waters of continental drainage systems. After several lines of divergence in fresh waters, the early vertebrates continued in these environments without entering into the seas. These waters might have received large quantity of detritus from the nearby lands, as many of the upper Silurian and Devonian animals were found in rocks containing sandstones and black shales. The huge placoderms which were not able to survive in the waters of rivers and small lakes, began to inhabit seas. The early sharks also shared the environment with these placoderms. However, the primitive bony fishes including the first actinopterygians were found only in fresh waters.

During Mississipian and Permian times, the lands were in depressed state on the earth with uniform climate, dense forests, swamps and topography. These environments might have favoured the adaptive radiation of amphibians. A change of environment like uplifts of continental areas, altered climatic conditions and varied topography of the land occured at the end of Pennsylvanian and early Permian periods. The streams, ponds and swamps changed into dry uplands and the animals experienced alternation of wet and dry seasons annually. When the reptiles became a dominant group at the beginning of Permian period, there was an extensive glaciation in the Southern hemisphere. During this time, the vertebrates were established on land, diversified into varied lines and distributed widely throughout the world. Thus at the end of Permian period, the land was dominated by cold blooded tetrapods namely the amphibians and reptiles. But the labyrinthodont amphibians were gradually giving way to the aggressive reptiles.

MESOZOIC ERA (THE ADVENT OF A NEW AGE)

This is 'the period of middle or medieval life' and it lasted for about 130 million years. This era evidences the dominance and extinction of ancient reptiles and so it is considered as 'the Golden Age of reptiles'. It is the time during which the birds and mammals have originated. Thus the Mesozoic era is the time of intermediate life' and it is divided into Triassic, Jurassic and Cretaceous periods. This era represents the time of 'middle life' in earth history.

Triassic Period

It is the earliest division of the Mesozoic era (181–230 million years ago) constituting a new phase of vertebrate evolutionary history. The Triassic lasted for 40 million years and was a transition period between the old life of Palaeozoic and the new life of Mesozoic times. The fishes, amphibians and reptiles progressed towards the future evolution of vertebrates. The labyrinthodont amphibians and mammal-like theriodonts (therapsids) developed advanced characters before they became extinct. In addition, new land-living vertebrates such as the ancestors of frogs, the first representatives of turtles as well as highly specialized marine reptiles like ichthyosaurs, placodonts, nothosaurs, plesiosaurs, etc., arose during Triassic period. Rhynchoccephalians were widely distributed in continental regions. Thecodonts, crocodiles, dinosaurs, ictidosaurs and the first mammals also appeared. The archosaurian theodonts were the first reptiles in this period and gave rise to the dominant reptilian groups of Mesozoic era namely the crocodiles, flying reptiles and dinosaurs.

Thus the Triassi is a period of transition in which various groups of tetrapods persisted from the Permian period, lived side by side with numerous new reptiles and evolved towards the dominant animal groups of middle and later Mesozoic era. The varied climatic and environmental conditions of Permian period was also the same for the Triassic so that the land vertebrates were found throughout the world. Moreover, the Triassic period enabled the land vertebrates to move from one region to another so that the amphibians and reptiles were widely distributed. In general, the domination of dinosaurs and pterosaurs, origin of first egg-laying mammals and extinction of the primitive amphibians were the important events during this period.

Jurassic Period

This period lasted for 30 million years dating back to 135–181 million years. The beginning of Jurassic was characterized by the spread of marine waters so that the lands became much restricted. As a result, the seas intruded into many continental regions. Therefore, the fossil record of land animals from lower Jurassic to middle Jurassic periods is scarce. In addition, the desert environments would have prevailed without supporting many animal groups. During the late Jurassic, broad low lands were present in various parts of the world and were most favourable for the dinosaurs. The abundant vegetation was used as food by the herbivorous dinosaurs upon which the large carnivorous dinosaurs preyed. Thus this period was an 'age of giants'. The first toothed birds appeared and were poor flyers as they were in early stage of divergence. They had a heavy competition with the flying reptiles. In addition, lizards, crocodiles and very ancient mammals also made their appearance during this time. It was the time in which there was a transitional evolutionary development from the reptilian to mammalian stage.

Cretaceous Period

This period was prevalent about 63–135 million years ago and lasted for 60 million years. The early Cretaceous period was also characterized by having the same environmental conditions of late Jurassic. Therefore, the land animals were much restricted but the early Cretaceous beds were found to contain dinosaurs. On the other hand, the fossil record of late Cretaceous period was extensive. The lands began to elevate with the extension of continents and had flowering plants and deciduous trees. Thus it was a time of modernization of plant world. The conditions were more favourable for the fossilization of land animals.

There were accumulation of marine deposits in an extensive scale and they were ample evidences for the presence of Cretaceous marine reptiles and also for the radiation of modern bony fishes which was one of the major events in vertebrate evolution during Cretaceous period.

At the end of Mesozoic era, the life of dinosaurs came to an end. These great reptiles including the marine forms such as ichthyosaurs, plesiosaurs and mosasaurs along with the flying reptiles (the pterosaurs) which were dominant on land for more than hundred million years failed to survive. The disappearance of the dominant Cretaceous reptiles was a sudden dramatic event. The toothed primitive birds also became extinct but the modern birds and archaic mammals made their appearance. The first mammals of Triassic times and the modern placentals of Cretaceous times remained as small groups during the middle and late Mesozoic era. At the advent of Palaeocene, the land was inhabited by numerous groups of mammals.

CENOZOIC ERA

It is the period of modern life and is considered as 'the age of mammals'. This era lasted for 75 million years in which the evolutionary history entered into 'the modern phase' with the transition from 'the age of reptiles' into 'the age of mammals'. During the early Cenozoic times, the continental areas were lifted to new heights with the formation of modern mountain systems followed by the worldwide climatic changes. These changes caused alterations in global plant life resulting in the evolution of flowering plants and grasses as new habitats. These events could have led to the extinction of the dominant reptilian groups and to the rise and radiation of the mammals. The Cenozoic era is divided into two periods namely Tertiary and Quarternary.

Tertiary Period

It lasted for about 74 million years and was characterized by the origin and evolution of mammals to a large extent. This period is split into five epochs.

Palaeocene epoch It lasted for 17 million years during which the placentals became abundant and radiated.

Eocene epoch It lasted for 19 million years and was characterized by the specialization and diversification of placentals, establishment of hoofed and carnivorous mammals and extinction of most of the primitive egg laying mammals.

Oligocene epoch It lasted for 11 million years and was the period of appearance of higher placentals and anthrapoid apes and extinction of the primitive mammals.

Miocene epoch It existed for about 16 million years during which the mammalian evolution reached the peak and the first man-like apes appeared.

Pliocene epoch The life span of this epoch was about 11 million years with the dominance of elephants, horses and camels. The most important event of this time was the origin of man from apes.

Quarternary Period

This period was characterized by the occurance of glaciations with duration of one million years. The most important event in this period was the origin of man so that the Quarternary period is described as the 'Age of Man'. It includes two epochs namely Pleistocene epoch and Recent epoch. It is believed that the end of the Ice Age represents Pleistocene and the post-glacial period corresponds to the Recent epoch.

Pleistocene epoch It began about 0.975 million years ago during which many mammals declined and the social life of man started.

Recent epoch It is considered as a distinct 'Psychozric era' representing the time of mental life and it started from 25000 years ago. The Recent epoch denotes the period of 'intellectual evolution' as human society with social life was established during this time.

The fossils of Cenozoic mammals are available in plenty when compared to the fossils of Palaeozoic and Mesozoic amphibians and reptiles. The evolution of Cenozoic mammals can be grouped into four major events.

The first one was the radiation of archaic fauna of Palaeocene and Eocene and thedominance of mammalian groups that failed to survive in later times.

The second event was the replacement of the primitive fauna by the ancestors of the modern mammals during late Eocene and Oligocene.

Thirdly, the mammalian fauna became modernized and evolved into the advanced and specialized mammals of the present times. The final stage was the establishment of the present day fauna during Pleistocene time and the extinction of many larger Pleistocene mammals.

The usual geological practice is to place the oldest period at the bottom with younger periods successively in a sequence. Thus, the geological time scale along with earth's divisions, age, duration and the faunal details are given in Table 1.

Table 3.1 Geological Time Scales showing the earth's divisions, age, duration and the evolutionary events.

ERA	Period	Epoch	Duration (million years)	Time from beginning of period (million years)	Major evolutionary events
	Quarternary	Recent	0.025	0.025	Age of man and the period of 'intellectual evolution' with the formation of human society.
		Pleistocene	1	1	Decline of great mammals and beginning of man's social life.
Cenozoic (Age of mammals)		Pliocene	11	12	Origin of man and dominance of elephants, horses and camels.
		Miocene	16	28	Climax of mammalian evolution and the appearance of first man-like apes.
	Tertiary	Oligocene	11	39	Appearance of higher mammals and anthropoid apes and extinction of primitive mammals.

ERA	Period	Epoch	Duration (million years)	Time from beginning of period (million years)	Major evolutionary events
		Eocene	19	58	Diversification of placental mammals, establishment of hoofed and carnivorous mammals and extinction of archaic egg-laying mammals.
		Palaeocene	17	75	Abundance and radiation of placental mammals.
	Cretaceous		60	135	Radiation of modern bony fishes, prevalence of marine reptiles, extinction of dinosaurs and toothed primitive birds, appearance of modern birds and primitive mammals, presence of numerous groups of mammals.
Mesozoic (Golden age of reptiles)					
	Jurassic		30	165	Dominance of dinosaurs, appearance of first toothed birds as well as lizards, crocodiles and archaic mammals. Time of transition from reptilian to mammalian condition.

	Triassic	40	205	Wide distribution of amphibians and reptiles, dominance of dinosaurs, origin of first egg-laying mammals, extinction of primitive amphibians, appearance of ancestral frogs, turtles, crocodiles and first mammals and wide distribution of Rhynchocephalians.
Palaeozoic (Age of ancient life)	Permian	25	230	Appearance of mammal-like reptiles and the extinction of several lines of amphibians.
	Carboniferous (Pennsylvanian and Mississipian)	50	280	Development of amphibians, appearance of stem reptiles and spread of ancient sharks and amphibians.
Palaeozoic (Age of ancient life)	Devonian	45	325	Abundance and domination of many species of fishes (Age of fishes) and appearance of first amphibians.
	Silurian	35	360	Appearance of wingless insects and marine arachnids and appearance of first fishes.

ERA	Period	Epoch	Duration (million years)	Time from beginning of period (million years)	Major evolutionary events
	Ordovician		65	425	Origin of fresh water fishes and presence of a number of invertebrate fossils.
	Cambrian		80	505	Presence of a number of marine invertebrates (Age of Invertebrates)
Proterozoic (Period of primitive life)			1500	2000	Presence of marine invertebrates as fossils and their loss due to changes on earth's crust.
Archaeozoic (Period of primal life)			1000	3000	Evolution of probable primordial heterotrophic organisms and absence of fossils.

BASIS FOR NAMING THE GEOLOGICAL TIMES

Various geological times are being named based on the early workers who studied the rock strata in geological science or based on the fossils that they discovered, as detailed in Table 2.

Table 3.2 Various geological times and the basis of naming

Era	Names of geological times	Based on/from
Palaeozoic era	Cambrian	Cambria, the ancient name of Wales
	Ordovician	Ancient tribe, the Ordovices lived in Southern England and Wales.
	Silurian	Ancient tribe, the Silures lived in Southern England and Wales.
	Devonian	English country of Devonshire.
	Mississippian and	The North American regions along with the Mississippi river
	Pennsylvanian	and the Alleghery Mountains of Pennsylvania.
	Carboniferous	The Carbonaceous rocks of many continental areas.
	Permian	The province of Northern Russia namely Perm.
Mesozoic era	Triassic	Subdivision of the time into three units in Central Europe during its first study.
	Jurassic	The Jura Mountains on the Alpine region.
	Cretaceous	Latin word 'Creta' means chalk.
Cenozoic era	Tertiary	Third great division in earth history.
	(The names of epochs are formulated by the English geologist, Sir Charles Lyell)	
	Palaeocene	…….. Ancient division of Cenozoic era.
	Eocene …….	Dawn of recent times.
	Oligocene	…….. 'Few recent time'
	Miocene ……..	'Less recent time'
	Pliocene …….	'More recent time'
	Pleistocecne	……… 'Most recent time'
	Quarternary …………..	Fourth great division in earth history.

THE SEQUENCE OF VERTEBRATE EVOLUTION THROUGH GEOLOGICAL TIMES

Though the first vertebrates were identified in Cambrian rocks, only very little fossil evidence is available. But the fossils of late Silurian give some more but scanty evidence for the early vertebrates. The actual vertebrate fossil record is found in Devonian rocks. This indicates the starting of the vertebrate evolution from the middle of the Palaeozoic era. The fossil record reveals that all major groups of fishes appeared during the middle and late Devonian times. The first land vertebrates namely the amphibians descended from certain advanced fishes during the transition from the Devonian to Mississippian periods. The reptiles evolved from amphibians in Pennsylvanian period. During Mesozoic era, dinosaurs were the dominant animals and were the rulers of the earth for more than 100 million years, but they all became extinct.

Prior to the disappearance of dinosaurs, two major vertebrate groups evolved from reptilian ancestors. One group was mammals which appeared during the Triassic and the another one was birds during the Jurassic. The birds became highly specialized by Cretaceous times and were widely distributed in continents and islands during early Cenozoic era. During the transition from Cretaceous to the beginning of Tertiary period, the opportunities were more to the early mammals (the beginning of the 'Age of Mammals') but it was the time for the decline of the reptiles. It was this period in which man, the intellect mammal evolved within a short period.

Recently a group of scientists at the symposium held at the British Geological Society, London, 2011 feel that the humans are changing the earth dramatically and *Homo sapiens* will define a geological period namely Anthrapocene. But they are not sure whether this 'age of man' will be short or long. For the first time in earth's history, a single species (man) has radically changed the earth's morphology, biology and chemistry and he is also well aware of the happenings. The scientists believe that the geological time is abruptly changing in the hands of humans and this is evidenced by the alteration of atmospheric composition due to excessive use of fossil fuels, increased carbon-di-oxide level in the lithosphere with the resultant global warming, massive melting of ice and snow on earth's crust, acidification of the oceans and so on.

DETERMINATION OF AGE OF EARTH

Several techniques have been put to use to determine the age of rocks and its fossils. The following are a few methods used to measure the age of earth.

1. ***Salinity method*** It is one of the methods of the past and is based on the rate of salinity of the oceans. According to this method, the annual increase in salinity of the sea water is taken into consideration for the estimation. But this method is not accepted because the consistency of increased rate of salinity in oceans is doubtful.

2. ***Rock sedimentation method*** By this method, the age of earth is estimated based on the rate of sedimentation of rocks. However, there is no continuous record of sedimentary rocks and so this method is also not accepted.

Radioactive method It is the most popular, accurate and recent method introduced by Boltwood (1907) to determine the age of rocks. It is based on the disintegration property of the radioactive elements. The minerals in the rocks alsocontain little quantity of radioactive substances such as radium, uranium and thorium. These materials undergo disintegration at a steady rate to become stable element, the Pb^{206} ($Lead^{206}$). The rate of this disintegration is not affected by temperature, pressure, chemical conditions and so on. One-half of the total number of $uranium^{238}$ atoms will break down to form lead and helium in a period of 4.5 billion years. Therefore, the age of the rock can be calculated by the accurate estimation of uranium and Pb^{206} in the rocks. By following the same method, radioactive Thorium ($Thorium^{232}$ into Pb^{206}), Rubidium ($Rubidium^{87}$ into $Strontium^{87}$), etc., are also used to estimate the age of earth.

However, this method is applicable only to the igneous rocks which contain radioactive substances and so it cannot be used in sedimentary rocks or in fossils.

Radioactive carbon method This method is based on the fact that all living organisms contain a heavy carbon isotope (C^{14}) which will disintegrate into C^{12} when the organism is dead and buried. Therefore, the age of fossil can be estimated by measuring the amount of C^{14} in the fossil through half-life method. But this procedure is useful only for those fossils which contain organic material and it cannot be used in oldest fossils as there is a steady decrease of C^{14} with time.

Potassium–argon method This is a recent technique introduced to estimate the age of earliest known hominid fossils. The ordinary Potassium (K) contains 0.01% radioactive isotopes which will disintegrate into Ca^{40} and $Argon^{40}$ with a half-life of 1.3×109 years. Thus the age of a fossil can be determined by half-life method.

QUESTIONS

1. Give an account of geological divisions and major events that occured in each division.
2. Describe various geological eras and periods and add a note on dominant vertebrates of geological times.
3. What is the basis for naming the different geological divisions?
4. Explain briefly the environmental conditions and vertebrate fauna during Palaeozoic/Cenozoic era.
5. Write a short note on the sequence of vertebrate evolution during geological times.
6. In what ways the age of earth could be determined?
7. Write notes on :
   a) Geological time
   b) Geological time scale
   c) Carboniferous period
   d) Devonian period
   e) Palaeozoic era
   f) Mesozoic era
   g) Cenozoic era

4

ORIGIN OF CHORDATES

Chordate evolution is a history of transformations and is marked by physical and behavioural specializations. These evolutionary modifications and specializations have led to the development of a variety of forms from a single ancestral group. The chordate taxa include two major subphyla namely Protochordata or Acraniata and Craniata or Vertebrata. The subphylum Protochordata comprises three major groups such as Cephalochordata, Hemichordata and Urochordata. The group Vertebrata includes two major divisions namely Agnatha or Monorhina and Gnathostomata or Amphirhina. The agnathans are primitive jawless vertebrates such as extinct ostracoderms and living cyclostomes and mixines whereas the gnathostomes include the primitive extinct armoured jawed fishes like placoderms and acanthodians as well as fishes and tetrapods like amphibians, reptiles, birds and mammals.

The chordates are deuterostomes and are characterized by having a notochord, a dorsal tubular nerve cord and pharyngeal gill slits (Figure 4.1).

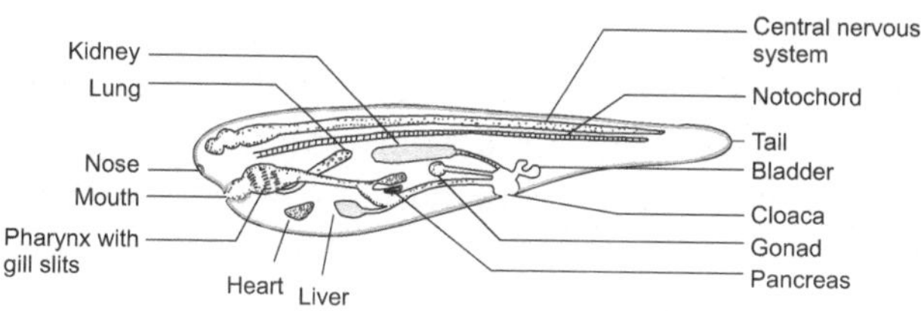

Figure 4.1 Generalized typical vertebrate animal

All vertebrates, in addition to the above chordate features, possess the following characteristics.

i. Brain enclosed in the skull (Craniata)
ii. Vertebral column replacing the notochord (Vertebrata)
iii. The endoskeleton comprising of the vertebral column, head skeleton (cranium), limb girdles and paired appendages
iv. Muscles attached to the skeleton to ease the movement
v. Muscular perforated pharynx
vi. Closed circulatory system with well-developed ventral muscular heart and blood with red blood corpuscles
vii. Hepatic portal system
viii. Post anal tail and
ix. Cephalization and bilateral symmetry.

Thus the craniates form a monophyletic group within Chordate phyla.

Most probably, the chordates might have originated prior to the beginning of the Palaeozoic era. The available Ostracoderm fossils clearly indicate that the chordate evolution might have taken place before the Ordovician and Silurian periods. In general, the chordate origin can be traced from the extinct and living invertebrates as well as from the larval stages of certain invertebrates. It seems to be true that almost every phylum in the animal kingdom shows affinity with chordates. The primitive chordate animal would have possessed the same structure like that of Amphioxus. Later fish-like early chordate might have evolved with lophophore mode of feeding. Therefore, the present day higher pharyngeal feeding chordates might have originated from the lophophore or filter feeding type of animals. There are many theories regarding the origion of chordates and they are given as follows.

1. *The Coelenterate Theory* According to this theory, the coelenterates were related to primitive chordates in having radial symmetry and sessile habits. But the coelenterates are diploblastic acoelomate invertebrates and so this theory is not accepted.

2. *The Nemertine Theory (Kofoid and Hubrecht, 1883)* This theory suggests that the chordates might have originated from marine nemertines because they resemble Balanoglossus in having elongated vermiform body, smooth skin with unicellular glands, terminal anus, proboscis and ectodermal nerve plexus. The nemertines also exhibit affinities with other vertebrates in having dorsal nerve cord comparable to the spinal cord, the cerebral and lateral nerves resembling vertebrate brain, the later line sense organs, the proboscis sheath above the gut comparable to the notochord and so on. However, these similarities are only very superficial and are not confirmed. Therefore, this theory is discarded.

3. *The Annulate Theory (Semper 1875)* This theory assumes that the annulate invertebrates like annelids and arthropods (metamerically segmented invertebrates) might have given rise to the segmented chordates. But there are many differences between these two groups. In annulates, the body segmentation is almost complete and all the organs are affected by it to some extent. In these invertebrates, the nerve cord is present below the gut. But the segmentation is superficial in chordates with the nerve cord dorsal to the gut. In both groups, the blood circulation is in opposite directions. Moreover, the annelids exhibit typical protosome type of developmental pattern like spiral cleavage, gastrulation by the immigration of cells and so on. But the chordates show typical deuterostome embryology such as radial cleavage, gastrulation by invagination, etc. In annulates, the coelom is schizocoel whereas it is enterocoel in early chordates. Thus the chordates differ from annulates in many respects and so the annulates cannot be considered as the ancestors of chordates.

4. *The Arachnid Theory (Patten, 1912)* This theory relates the earliest fossil vertebrate ostracoderms with the primitive fossil arachnids namely Eurypterids of the Cambrian and Silurian periods. But these resemblances are only superficial so that this theory is also not accepted.

5. *The Echinoderm Theory (Bateson and Garstang, 1886)* This theory proposes the origin of chordates from echinoderms because of more close resemblance of echinoderms and their allies with chordates. The relationship between the echinoderms and early chordates can be known from the following similarities between the two groups.

- In both, the early development namely cleavage, gastrulation and mesoderm formation is similar.

- In both groups, the coelomic sacs are separate.
- The prochordate larva and the auricularia larva of echinoderms are similar. The most striking similarity in these larvae is the arrangement of longitudinal ciliary bands.
- The arrangement of nervous system is similar in the larvae of both groups.
- The serological studies indicate a close relationship between the two groups.
- Above all the biochemical tests reveal that the hemichordates and echinoderms possess both creatine phosphate and arginine phosphate which are involved in muscle contraction.

In addition, there are striking resemblances between the Tornaria larvae of hemichordates and the echinoderm larvae. The ciliated bands of bipinnaria and auricularia larvae are similar to the tornaria larva of Balanoglossus. In both, the coelom is enterocoel and the nervous system is in the form a network of nerve fibres. In both the larvae, the alimentary canal and the dorsal pore are similar.

The axial gland of chinoderm larvae is homologous to the glomerulus of Tornaria larva. In both cases, the anus develops at the closed blastopore. However, the Tornaria larva differs from the echinoderm larvae in some respects so that it is persumed that the two groups would have evolved from a common ancestor.

The above morphological, embryological and biochemical evidences suggest that the echinoderms and chordates are very closely related. But the echinoderm mentioned here does not mean the present day genera but the extinct Cystinoids with lophophore feeding mechanism.

Cephalodiscus (Figure 4.2) is the most primitive early chordate and is an intermediate form between the echinoderms and chordates in possessing lophophore type of feeding mechanism as well as gill slits.

Figure 4.2 Cephalodiscus

This early chordate indicates that a lophophore feeding animal could have converted into a metameric fish-like individual with its pharynx so that the pharyngeal mechanism could have substituted for the lophophore feeding. In this process, they might have lost the tentacles but developed the endostyle, gill slits and atrium.

6. *The Auricularia Theory (Garstang, 1894)* Garstang suggested that a ciliated auricularia-like larva could be converted into a fish-like animal with muscular locomotion and the adults could have substituted gill slits and endostyle for the original lophophore. According to this theory, the fish-like form arose by the development of muscles along the sides of the elongated body of the auricularia larva for muscular locomotion. The ciliated bands pushed upwards and eventually rolled up to form the neural tube. The adoral ciliated band might be modified into the endostyle (Figure 4.3).

Thus it is regarded that the organization of the adult tunicates could be derived from the sessible lophophore-feeding animal and the larval tunicates might be evolved from an echinoderm-like auricularia larva. The elimination of sea-squirt stage in vertebrates could be explained on the assumption that the early free swimming chordate would have originated from some ascidian tadpole-like larva by paedomorphosis or neoteny.

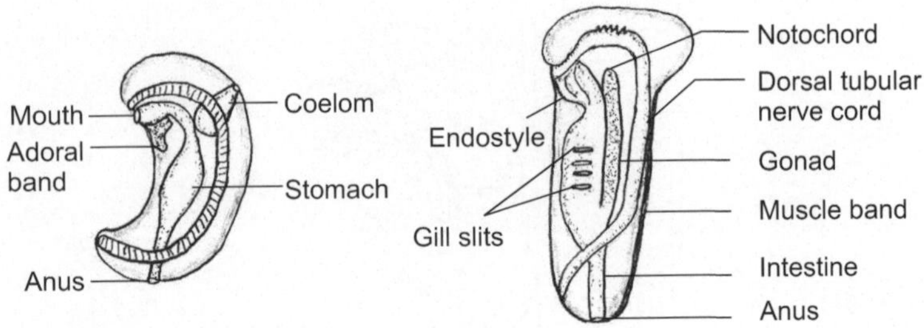

Figure 4.3 Comparison of Auricularia larva with Prochordate larva

7. ***Bateson and Garstang Theory*** Garstang's view has been strongly supported by Bateson and they believed that the chordates are related to the sessile lophophore-feeding animals rather than the annulates. These lophophore feeders would have acquired a pharynx with gill slits and the larvae developed muscles, notochord and a nerve tube. Then the sessile phase might have disappeared with the beginning of the course of evolution of free chordates.

The deuterostomes which include vertebrates, could have originated from sessile protosomes with lophophore type of feeding. The echinoderms are the major group of animals which are very close relatives to the chordates and hemichordates, thus evidencing that Deuterostomia is a monophyletic group. The urochordates and cephalochordates are the two basic groups of living chordates. *Ciona*, a tunicate possesses similar structures like those of cephalochordates and the larvae of cyclostomes. The hemichordates are the living marine deuterostomes of which two groups namely Pterobranchs (e.g., *Cephalodiscus*) and Enteropneusts (e.g., *Saccoglossus*) are more closely related to the echinoderms than to the chordates. Thus it is probable that the echinoderms were the early offshoots and the hemichordates were the next to be evolved. However, the Pterobranchs possessing lophophore (hemichordates) could be the earlier offshoots than the echinoderms. According to Chaeffer (1987), the hemichordates constitute a sister group of echinoderms.

From the foregoing account, it is clear that the echinoderms and lower chordates exhibit striking similarities even in their development. Thus this phylogenetic relationship is unquestionable. Therefore, it appears reasonable that the chordate stock arose from animals with the chordate-filter feeding system. The auricularia-like larva might be the ancestors to the ascidian tadpole larva-like chordates. Some of them could have undergone retrogressive metamorphosis resulting in adult sessile ascidians and some others could have developed into fish-like animals by neoteny. According to Haeckel (1866), the tunicates were more evolved than the vertebrates, recapitulating their vertebrate heritage in the form of larval stage. Amphioxus-like animals could have been derived through readaptation from an individual that was relatedto the simpler appendicularians. Berril (1955) ascertained that Amphioxus and vertebrates would have evolved from ascidian-like stock through Paedomorphosis. According to him, the hemichordates played no direct role on chordate evolution and the Amphioxus is a degenerate rather than a primitive form. It is also considered that the echinoderms and hemichordates could have evolved from a common ancestor. Bone (1960) has opined that the urochordates, cephalochordates and chordates would have evolved from the hypothetical 'Protohemichordate ancestor'.

The earliest vertebrate fossils are jawless Ostracoderms which had many characteristic features similar to the living vertebrates. These early vertebrates would have evolved from an invertebrate–chordate lineage. A mid-Cambrian invertebrate–chordate fossil namely *Pikaia* resembles *Amphioxus* in having a ribbon-shaped fish-like body with a length of 5 cm, notochord and V-shaped myomeres. However, it is suggested that *Pikaia* could be an early Cephalochordate. Any how, the tunicates seem to be in the line of most basal chordates but no definite tunicate fossils are available from Cambrian period.

Thus inspite of all these views, the origin of chordates is still a matter of speculations.

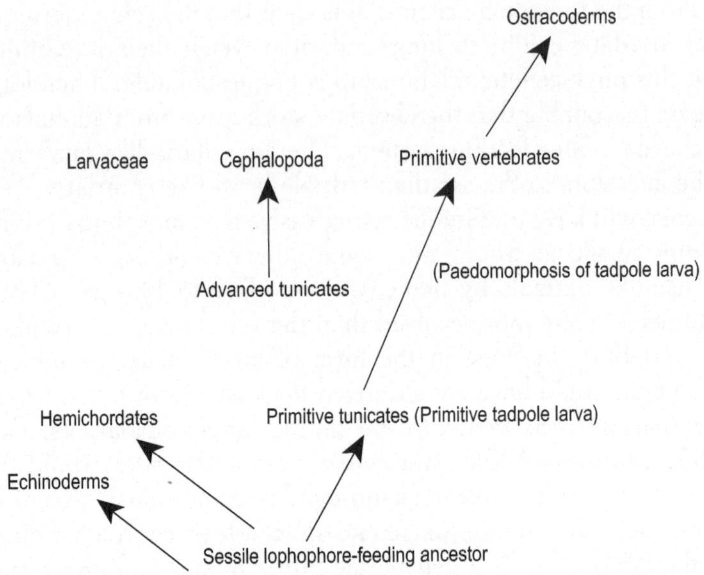

Origin of Chordates

QUESTIONS

1. Write an essay on the origin of chordates.
2. Give an account of various theories regarding the origin of chordates.
3. In what ways Bateson and Garstang theory is useful in tracing the evolution of chordates?
4. The origin of chordates is still a matter of speculations—why?
5. Write notes on the following:
   a) The Annulate Theory
   b) The Echinoderm Theory
   c) The Auricularia Theory
   d) Paedomorphosis and chordate evolution
   e) Bateson and Garstang theory

5

EVOLUTION OF OSTRACODERMS (Agnatha—Jawless Vertebrates)

The earliest known vertebrates are included in the group Agnatha (jawless vertebrates) and were known to appear about 550 million years ago. They include all the extinct and extant forms, the ancestors of which were a very successful group in Palaeozoic seas about 570–245 million years ago. While many of them resembled the modern hag fishes and lampreys, others had bony armour plates for protection with a length of 5 feet. The fossil hag fishes and lampreys date back to 350 million years. The best known ancient agnathans are the ostracoderms among which the members of Conodontia were the earliest forms. These were very small ostracoderms (less than 3 inches in length) without internal skeleton but with teeth, tongue and caudal fin. The conodonts first appeared in the fossil record about 515 million years ago and disappeared around 205 million years ago. Two other ancient groups of jawless vertebrates were Arandaspida (495 million years ago) and Astraspida (450 million years ago) which became extinct by about 440 milllion years ago. Later four groups of typical ostracoderms evolved and the earliest of these were Heterostraci followed by Pteraspidomorphs. The Pteraspidomorphs were replaced by the Cephalospidomorphs of which Anaspida and Galeaspida evolved first followed by Osteostraci. These ostracoderms appeared 430–425 million years ago and all became extinct.

This splitting up of ostracoderms into a number of diversified groups clearly evidences that they might have existed for a long time before they became extinct. The ancestors of these forms would have had little or no bony parts so that they were not fossilized. A recent study reveals that the ostracoderms like anaspids, pteraspids and Cephalaspids are in line with the descent of true fishes, hag fishes and lampreys respectively. Cephalaspids and its relatives exhibit many similarities to the lampreys and they are found to have the same ancestral stock. Therefore, it is probable that the true fishes would have been derived from the generalized anaspid stem.

The fossil record of ostracoderms is restricted to the late Silurian period through Devonian and they might have originated in Cambrian times from some unarmoured ancestors. In the rocks of middle Silurian, complete skeletons of two fossil forms namely Jamoytius and Thelodus were found. These would have evolved from the unarmoured primitive chordates which gave rise to the ostracoderms. During Devonian period, they became adapted to various modes of life and widely diversified as we know at present. But they disappeared completely at the end of Devonian except a few representatives like lampreys and hag fishes. This is because they were unable to compete with the jawed vertebrates which evolved during that period. However, the fossil record shows no link between the ostracoderms and higher jawed fishes so that information with regard to the ancestors of the jawed vertebrates are not available.

The ostracoderms including the modern living forms are included in orders namely Cephalaspida or Osteostraci, Anaspida, Cyclostomata, Pteraspida or Heterostraci and Coelolepida. The representatives of the above groups possess some general characters like absence of jaws, paired fins and bony vertebral column and the presence of a single pair of fins behind the head and well developed armour of bony plates or scales. Moreover, they had no internal skeleton so that they might be cartilagenous. That is why they were not preserved. Except for these general resemblances, all the members of ostracoderms were very different from each other, thus evidencing their independent lines of evolution.

Order : Cephalaspida (Osteostraci)

Example: *Cephalapsis* (Figure 5.1) and *Hemicyclapsis* from upper Silurian and Devonian

- These were small fish-like, heavily armoured animals.
- The head was protected by a strong and solid shield and the flat under surface of the body byvertically arranged, elongated bony plates with side to side flexibility.
- The long body had a small median fin on dorsal side, paired lateral fins at the corners of the headshield and a terminal heterocercal tail fin.
- Paired eyes were closely placed on the dorsal side of the head with a single nasal opening inbetween and slightly in front of the eyes.
- A well developed pineal opening was present immediately behind the nostrils in between theeyes.
- Three sensory fields of some sort were present on the head shield in th form of depressed areascovered by small polygonal plates.
- The mouth was ventral in position and a series of small marginal gill openings were present.
- The brain was of a primitive type and the ears had only two semicircular canals.

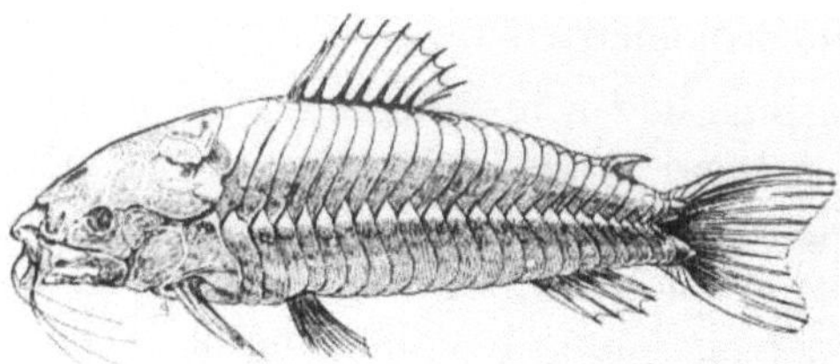

Figure 5.1 *Cephalapsis*

The head shields of many Cephalaspids were found to vary widely. The body plan of these animals clearly shows that they could be the bottom-living vertebrates found in shallow waters of streams, lakes and estuaries, feeding on small food particles by sucking through the mouth.

Order : Anaspida

Example : *Jamoytius* and *Thelodus* from Silurian and *Birkenia* from upper Silurian

This group resembles the cephalaspids in having a body covered with bony plates or scales, laterally placed eyes, a single nostril, a small pineal opening and marginal gill apertures. But these were very tiny animals with narrow and deep body adapted for active swimming. They had the following characteristics :

- The head was covered with small scales which were arranged in complex pattern.
- The mouth was terminal in the form of transverse slit without jaws.
- The body scales or plates were vertically elongated and were arranged in longitudinal rows.
- The paired fins were absent but small spines were present on each side of the body.
- On the back of the body, many median dorsal fins were present.
- The anal fin was well developed with a large lower lobe and a smaller upper lobe (reversedheterocercaltype) which could have been advantageous for active swimming in these heavilyarmoured animals. Though they swam near the surface waters feeding on planktons, they werenot very efficient swimmers.

Some anaspids had a marked reduction in the body armour, thus showing a trend towards the direction of modern lampreys. *Jamoytius* (Figure 5.2) was a very primitive jawless vertebrate with small elongated and tubular body having terminal suctorial mouth, circular gill openings on each side of the head, lateral fin folds, a long dorsal fin and a tail fin with a long lower lobe and shorter upper lobe. These animals seem to occupy a position close to the ancestors of the lampreys and their relatives.

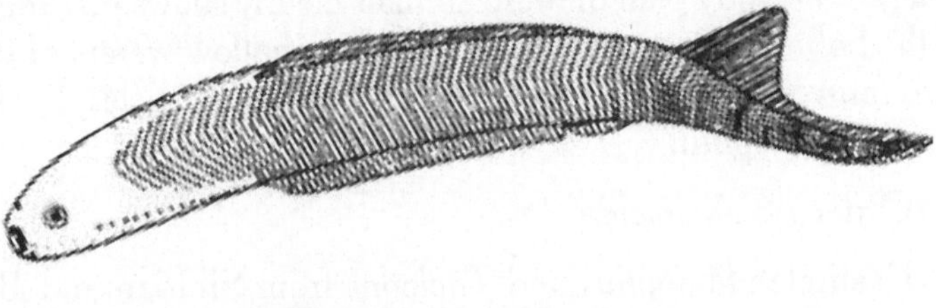

Figure 5.2 *Jamoytius*

Order : Pteraspida (Heterostraci)

Example *Pteraspis* (Figure 5.3) from Silurian and Devonian and *Drepanaspis* from lower Permian

These were the earliest known ostracoderms with well developed body armour but without true bones. They exhibit a varied range of adaptive radiation. For example, some like Pteraspis had the following features:

- These were small and heavily armoured ostracoderms.
- They had heavy head shield, the lower surface of which contained transverse slit-like terminal mouth.

Figure 5.3 *Pteraspis*

- The eyes were lateral in position on the head with a pineal opening at the top but without visible nasal opening.
- Gills had single exit on each side of the head shield.
- Median and paired fins were absent but reversed heterocercal tail fin was present. They were probablyactive swimmers in surface waters.

  Others like Drepanaspis had the following features :
- These were the largest bottom-living ostracoderms with flat and broad body.
- The head shield was in the form of a series of plates of varying sizes.
- The eyes were found on the sides of the head and were widely separated.
- The wide mouth was situated on the front of the shield.
- They had a small tail.

Order : Ceololepida

Example : *Thelodus* (Figure 5.4) and from upper Silurian and lower Devonian

These are considered to be specialized ostracoderms and are known from their impressions in rocks. Though very little information is known about the members of this group, it is evidenced that they had flattened body with laterally placed eyes. The coelolepids differ from other ostracoderms in possessing a body covering formed off by minute denticles. They had heterocercal tail fin.

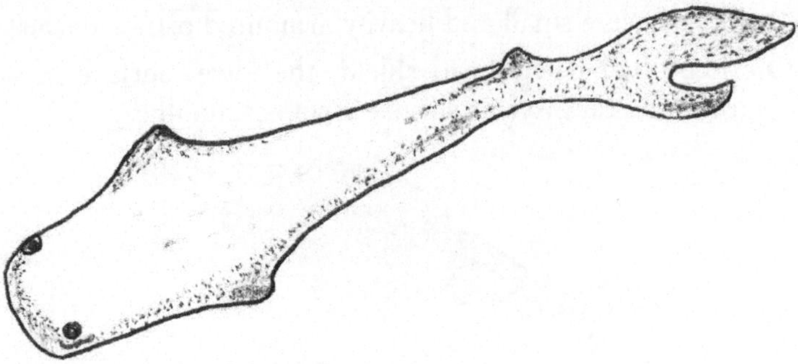

Figure 5.4 *Thelodus*

Order : Cyclostomata

Example : Modern lampreys (Figure 5.5) and hag fishes

The cyclostomes are highly modified and degenerate survivors of some of the first vertebrates that appeared in the fossil record. The lampreys closely resemble some of the early fossil vertebrates. They have eel-like naked body with a long fin on the back and around the tail. The eyes are lateral in position and the nostrils are situated between the eyes on the top of the head. The mouth is round and is provided with sharp teeth. The young ones are bottom-living whereas the adults lead a parasitic life sucking the blood of fishes. Eventhough the lampreys possess specialized characters, they are primitive in many respects. The cyclostomes include two groups namely Myxinoidea (hag fishes) and Petromyzontia (lampreys) which are not closely related. But they are considered to be the end products of two lines of evolution from two different groups of ostracoderms. The former group is regarded as the degenerate offshoot of Pteraspida whereas the latter one is a slightly less degenerate descendant of Cephalaspida. Anyhow, the cyclostomes are the most primitive of surviving vertebrates.

Figure 5.5 Lamprey (*Petromyzon*)

QUESTIONS

1. Explain the role of ostracoderms in the evolution of vertebrates.
2. Give an account of different groups of ostracoderms and their evolutionary significance.
3. Write an essay on Agnatha and their importance.
4. Write notes on the following :
   a) Fossil ostracoderms
   b) *Cephalapsis*
   c) *Jamoytius*
   d) *Pteraspis*
   e) Cyclostomes

6

EVOLUTION OF PRIMITIVE JAWED VERTEBRATES (Acanthodians and Placoderms)

Acanthodians are the earliest known small vertebrates with true jaws. They have the resemblance to that of sharks so that they are given the name 'spiny sharks'. These were the primitive and the first known fresh water gnathostomes in late Silurian period. They were found in rivers, lakes and swamps of middle and late palaeozoic times. They were dominant during early Devonian and became extinct at the end of Palaeozoic era. Placoderms are also primitive gnathostomes which evolved through a wide varied divergence during middle Palaeozoic times. They represent a heterogeneous group that appeared along different lines with those of acanthodians. Most of the placoderms became extinct by the end of Devonian and all disappeared at the end of Palaeozoic era around 360 million years ago. The exact ancestors of placoderms are not clear but it is evident that they might have evolved from very primitive gnathostomes.

ACANTHODIANS

The oldest members of acanthodians are known only by the scales and spines found in the Silurian rocks as complete skeletons of acanthodians are not available. They had primitive upper and lower jaws as in the genus *Acanthodes* from middle and late Palaeozoic periods. These fishes had enlarged upper jaws and well developed lower jaws, both bearing sharp teeth. Behind the jaws, there were gill arches (hyoid arches) in which the upper hyomandibular bone was enlarged in order to connect the jaws to the skull.

An early acanthodian namely *Climatius* (Figure 6.1) from late Silurian and early Devonian times represents a generalized good example for the ancient gnathostomes. These were small vertebrates measuring a few inches in length with a fish-like body which was tapering from anterior to posterior end. The body was covered by a dermal armour of small scales. These scales were in the form of regularly arranged plates in the head region. The eyes were large and were placed very anteriorly so that the nasal region was very much restricted in the front end of the head. The upper jaw was formed by three pieces without teeth whereas the lower jaw had well developed teeth. There were five gill arches on either side of the head covered by opercular flaps somewhat similar to that of higher fishes. They had two large triangular fins on the back along with a large ventral anal fin, paired pectoral fins behind the skull and paired pelvic fins in front of the anal fin. The caudal fins were heterocercal type. In addition, five pairs of smaller fins were found on the ventral side of the body in between the pectoral and pelvic fins. These fins were found to be the characteristic of all acanthodians. All these fins were privided with spines, the significance of which was not known.

Figure 6.1 *Climatius*

However, the later acanthodians evolved with marked variations through Palaeozoic era until the close of the Permian period. During this period, some of them became long and slender and others evolved into deep bodied acanthodians. In some representatives, there was loss of both median and paired fins.

PLACODERMS

The placoderms were more specialized than the acanthodians. They had certain common basic characteristics such as strong upper jaws fused with the skull and lower jaws, head covered by dermal bones, gills situated far forward beneath the skull, presence of paired fins and so on. The placoderms were the highly diversified group with varied adaptations. Therefore, it may be difficult to classify them and to expect more generalized representatives. In general, the placoderms are included in the following orders.

1. ***Arthrodira*** The arthrodires are commonly called as 'armoured fishes' or 'joint - necked fishes' and are most varied well known placoderms of Devonian period. They were dominant during late Devonian and became extinct at the end.

Coccosteus (Figure 6.2) was a small fish-like typical arthrodire measuring about one to two feet in length. The front portion of the head and shoulder region were heavily armoured but the posterior of the body is completely naked. The skull was quite deep and strongly arched with a series of bony plates. The eyes were large and situated near the front of the skull. But the nostrils were small and were found at the very front end of the skull. The lower jaws were very strong and were attached to the back of the skull. A marked gap was present dorsally between the back of the skull and the front part of the body armour. This feature might have facilitated a strong upward movements of the head in these animals. The mouth was very wide as it was opened by jaws as well as by raising the skull. The operculum was absent so that the gills were present on either side of skull in the cheek region. A well developed cartilageneous notochord bordered by spines, a median dorsal fin and a caudal fin of heterocercal type were present.

Figure 6.2 *Coccosteus*

The earlier arthrodires were small, flattened and bottom-living animals competing with the early ostracoderms. Then they began to evolve towards increased size and mobility. They became fast swimming wild predators as in the case of *Coccosteus* during the middle and late Devonian times. But some like *Dinichthys* and *Titanichthys* appeared in upper Devonian with large skulls and strong jaws having large cutting plates. Dinichthys was considered to be the dominant genus with a body length of thirty feet.

2. *Petalichthyida* This group includes restricted specialized placoderms represented by *Macropetalichthyes* and *Lunaspis*. These resembled the arthrodires in having strong bony head shield, similar brain, etc., Though these were related to the arthrodires, they might have diverged at an early stage in separate evolutionary line.

3. *Antiarchi* This order comprises small, bottom-living and heavily armoured placoderms like *Bothriolepis* (Figure 6.3) and Pterichthys known from upper Devonian rocks. In Bothriolepis, the head was covered by a short shield formed by large plates and the body shield was long and box-like. The eyes were dorsally placed very close to each other in between the pineal opening. The mouth was small and ventral in position with very weak lower jaws. Thus, Bothriolepis resembles some of the ostracoderms superficially. At the same time, it also had well developed lungs opening from the pharynx. This was a common characteristic found in the early vertebrates of Devonian period during major lines of fish evolution.

Figure 6.3 *Bothriolepis*

In general, the antiarchians had long pointed, armoured and freely movable pectoral appendages attached to the front portion of the body shield. The body was tappering posteriorly forming a sort of heterocercal tail fin. In this way, the general plan of this group is similar to the arthrodires. But both groups would have evolved from a common ancestor and the antiarchi vertebrates might have assumed a trend towards bottom-living habitat.

4. *Rhenanida* The rhenanid placoderms are otherwise known as the stegoselachians as they resembled sharks in certain characters. So, it was considered that these ancient Devonian placoderms were the early relatives of the sharks. The genus *Gemuendina* (Figure 6.4) is a typical best known stegoselachian. It had a broad head and tapering flat body covered with small tubercles similar to that of the denticles of the modern sharks. The mouth was terminal provided with pointed teeth. The eyes and nostrils were situated on the dorsal side of the head. The presence of enlarged paired pectoral fins led to resemble remarkably the modern rays. But the jaws were of the primitive placoderm type. The similarity of these animals to the sharks and rays suggests that the two groups could have evolved along similar lines as two separate groups. Later the bottom-living skates and rays might have followed the mode of life of the rhenanids of Devonian times, thus showing a best example for convergent evolution.

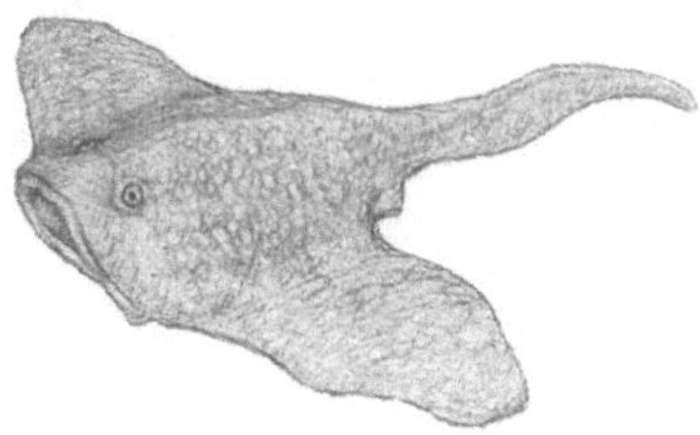

Figure 6.4 *Gemuendina*

Prior to the extinction of the placoderms during the transition between Devonian and Mississippian times, a number of *aberrant placoderm groups* evolved in separate lines. For example, the *Ptyctododonts* represented by *Rhamphodopsis* were small placoderms with reduced head and thoracic armour. The heavy dental plates of these animals indicate that they were adapted to feed on molluscs. Another group namely Phyllolepida which comprises flat, medium-sized and bottom-living genus *Phyllolepis* was considered to be the degenerate arthrodires. The genus *Palaeospondylus* was included in a placoderm group namely *Palaeospondyloidea* for the sake of convenience. This is because it is very difficult to interpret the details of their characteristics and relationships with other groups of placoderms.

QUESTIONS

1. Give an account of acanthodians.
2. Explain the characteristic features of various groups of placoderms.
3. Write notes on the following :
   a) *Climatius* b) Arthrodires
   c) Placoderms d) Acanthodians

7

EVOLUTION OF FISHES

The evolution of early fishes seems to have taken place during the Ordovician and Silurian periods of Palaeozoic era about 443–417 million years ago and were abundant in the Devonian period about 417–354 million years ago. During these periods, the fresh waters and seas were dominated by heavily armoured primitive fishes which gave rise to the modern sharks and bony fishes. The fossil fishes are not available in Ordovician until many early Ordovician astraspids and arandaspids were found. The late Ordivician fishes include a number of groups like thelodonts, shark-like fishes possibly acanthodians, conodonts and osteostracans. During Silurian, they were more abundant and diversified along with the radiation of the armoured, jawless ostracoderms and acanthodians. In Devonian times, all major groups of ostracoderms, the first sharks and six important groups of bony fishes and their derivatives evolved. The first extinction of fishes was recorded at the end of the early Devonian with the loss of Cyathaspids, acanthothoracids and others. A group comprising some agnathans, placoderms and some acanthodians disappeared during mid-Devonian. Within a time span of 15 million years, further fish groups including most agnathans, placoderms, remaining acanthodians and some bony fishes disappeared during late Devonian about 377–362 million years ago. The extinction of fishes seems to have been spread through the late Devonian so that only 17 families survived. The Devonian fish fauna were replaced by new groups of Carboniferous period.

At the expense of many groups of jawless fishes namely, placoderms and acanthodians, two major groups of fishes namely Chondrichthyes (sharks and rays) and Osteichthyes (bony fishes) evolved, diversified several times and became the marine and fresh water fish life of present day. These fishes appeared in Devonian times and might have originated in the Silurian period. Several Silurian and Devonian primitive fish lineages such as agnathans, lung fishes and coelocanths lived about 360 million years ago since the beginning of Carboniferous period. More recent bony fishes seem to have vast radiation over the past 150 million years. Both Cartilagenous and bony fishes could have evolved at the same time from different ancestors. The bony fishes were the most successful group of which chondrostei were the first bony fishes arising about 410 million years ago followed by Dipnoi (lung fishes) at about 407 million years ago. Then Coelocanths, osteolepiforms and Panderichyida appeared but they became extinct about 250 million years ago. Majority of the modern fishes might have evolved from the ancestors of chondrostei. These fishes appeared in fossil record only in late Permian times.

The appearance and diversification of sharks and bony fishes could be the probable cause for the disappearance of the ostracoderms, acanthodians and placoderms. From the beginning, the sharks and bony fishes are highly adapted for aquatic life, especially for active swimming. Thus they are good swimmers so that the early sharks and bony fishes succeeded over the primitive vertebrates. The body of the sharks and bony fishes has streamlined body for the rapid movement in water. The wedge-shaped head helps to split the water. The decreased body height, width and girth from front to back makes these fishes to glide in water with minimum resistance. In these modern fishes, the paired pectoral and pelvic fins are highly mobile and act as planes or elevators or rudders or breaks, thus playing a major role in controlling movement in water. The median dorsal and ventral fins along with the caudal fin act as stabilizers during locomotion in water. Moreover, the large caudal fin is very efficient for the forward movement of the fish in the aquatic medium. These fishes also possess efficient feeding, protective and reproductive mechanisms. Thus the modern fishes are well adapted for active aquatic life, most numerous and highly successful vertebrates.

MAJOR EVOLUTIONARY TRENDS IN THE EVOLUTION OF FISHES

During the evolution of fishes, there were major evolutionary changes which include evolution of jaws, teeth and paired fins as well as modification of hyoid arch.

Evolution of Jaws

The first vertebrates were the armoured jawless agnathans (ostracoderms) which lived in Devonian and Silurian times. Next, the acanthodians appeared in the middle Silurian and they were also armoured but had jaws. Then the jawed vertebrates namely the placoderms evolved in the middle to late Silurian times.

In the evolutionary history of the vertebrates, the development of movable jaws supported by skeleton with the ability to bite is a major turning point and appears as one of the greatest events.

The jaws were found to be originally derived from the gill arches. The presence of a large number of gills (as many as ten) is a characteristic typical feature of the primitive vertebrates like ostracoderms. At an early stage of evolutionary process, the first two anterior gill arches disappeared and the third one became changed into a pair of jaws. This view has been supported by many evidences especially by the study of development of certain modern fishes.

For example, in sharks, the arrangement of nerves in the head region indicates that the jaws are in series with the gill apparatus. A number of primitive jawed vertebrates possessed jaws in series with the gill arches (Figure 7.1).

The spiny sharks namely the acanthodians clearly indicate the period of entry of jaws in vertebrate evolution (late Silurian times).

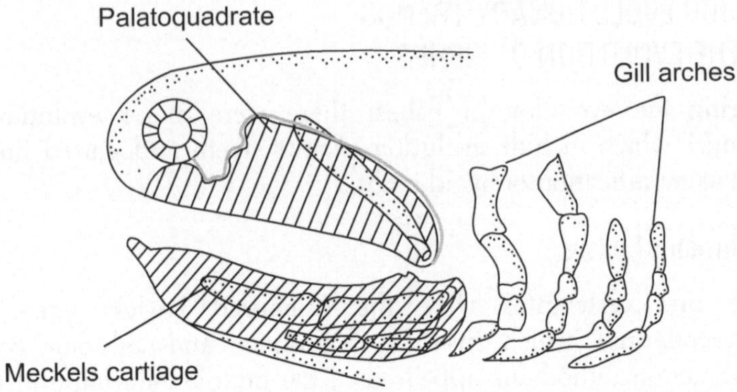

Figure 7.1 The head of primitive vertebrate (Acanthodian fish) with gill region

Modification in Hyoid Arch

In gnathostomes, the first arch is mandibular arch consisting of a dorsal palatoquadrate and a ventral Meckel's cartilage. Along the margin of the mouth, the palatoquadrate grows forward on two sides as upper jaw and the Meckel's cartilage develops as the lower jaw (Figure 7.2).

In the most primitive gnathostomes (primitive jawed fishes), the upper jaws were directly articulated with the cranium (autostylic jaw suspension). In the early period of fish evolution, the upper bone of the hyoid arch behind the jaws was transformed into a hyomandibular bone which was a connecting piece between the jaws and cranium (amphistylic jaw suspension). The fossil *Cladoselache* had amphistylic type of jaw articulation which seems to be primitive mode of suspension in jawed fishes. But in later sharks and many bony fishes, the upper jaws articulate with the skull only through hyomandibular piece (hyostylic jaw suspension).

Here, the gill slit that was originally the space between the cranium and hyoid arch was greatly restricted. This reduced gill opening remains as the spiracle in more primitive modern fishes but it disappears completely in highly advanced fishes.

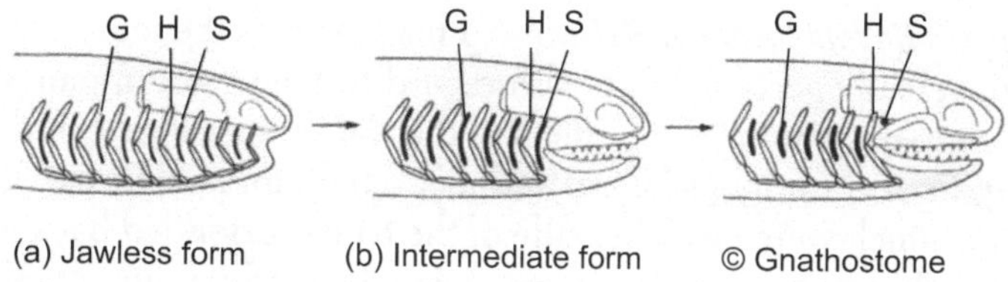

Figure 7.2 Diagrams showing the evolution of jaws in fishes

Evolution of Teeth

The fossil evidences clearly show that the teeth appeared at first in the lower jaw of *Climatius* (acanthodian) and the placoderms had teeth in both jaws. It is believed that the teeth might have derived probably from the dermal denticles of the ostracoderms. This is because the structure of the denticles is comparable to the teeth of modern sharks. It is assumed that the denticles, after development, would have migrated to the jaws assuming a new function. This is supported by the fact that the structure of placoid scales of sharks resemble the teeth of vertebrates.

Evolution of Paired Fins

Fishes have paired pectoral and pelvic fins in addition to unpaired median fins (dorsal, ventral and caudal fins). Various theories have been put forward by different authors with regard to the origin and evolution of fins in fishes and the following are the important ones.

Gill arch theory by Gegenbaur According to this theory, two pairs of posterior gill arches behind the head would have undergone modifications in their skeletal elements to form pectoral and pelvic fins. But since this theory has no supporting evidences it is discarded.

External gill theory by Graham Kerr This theory suggests that the paired fins along with their skeletal elements could have developed from the external gills of some larval fishes. This theory has also been discarded due to lack of evidences.

Ostracoderm theory by Gregory According to this concept, the paired fins and their skeleton must have derived from the body scales of ostracoderms at the posterior region of the skull. As this theory has no evidences, it is also not accepted.

Continuous fin fold theory by Balfour This theory is the most probable and universally accepted one with regard to the origin of paired fins in fishes. According to this theory, the ancestral fishes could have had a pair of continuous lateral and median fin folds. The two lateral fin folds which were on either side of the trunk, extended backwards, united behind the anus to form a single median fin, continued around the tail and ran along the mid-dorsal line. By breaking up of these continuous fin folds into a series of small fins, the median and paired fins might have developed (Figure 7.3). This theory has been supported by a number of evidences. In *Amphioxus*, the metapleural folds are continuous with the median fin on ventral side and are supposed to be the precursors of the paired fins. Primitive sharks like *Climatius* had seven pairs of fins in a continuous series. In *Cladoselache*, the pectoral and pelvic fins are nothing but lateral folds with unconstricted base. Skin folds are found to appear during the embryonic condition of some sharks. The tail fin of *Pleuracanthus* is in the form of most primitive type (diphycercal) as in Amphioxus. Thus Balfour's theory is supported by a number of facts and it has been widely accepted.

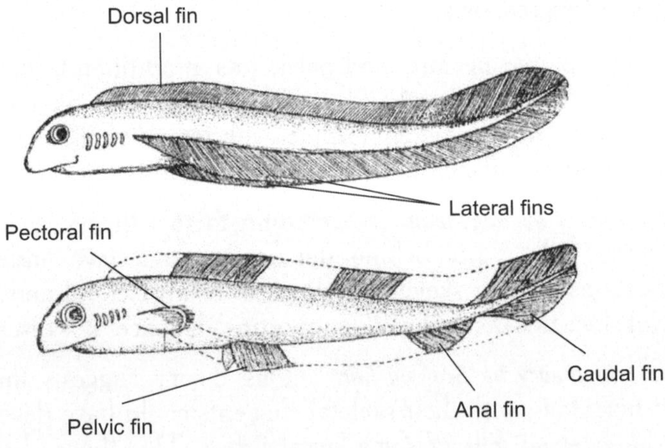

Figure 7.3 Origin of paired fins from continuous fin folds

EVOLUTION OF CARTILAGENOUS FISHES-SHARKS (CHONDRICHTHYES)

The sharks and the allied cartilagenous fishes are still dominating in the seas and they exist in a great variety of sizes and forms. Most of these fishes are not preserved as fossils because of their cartilagenous skeleton. But the teeth of these fishes were well fossilized. However, some shark-like fishes such as *Cladoselache*, *Climatius*, *Pleuracanthus*, etc., were well preserved. The *acanthodians* were the earliest shark-like fishes that lived in the rivers along with the ostracoderms. These fishes underwent extensive migration from the rivers into the seas during the end of the Devonian period. Though a few primitive fishes continued to live in rivers up to the end of Palaeozoic period, the entire group of sharks became exclusively marine at a relatively early time. Thus the sharks are not found in fresh waters at present. The sharks evolved in Devonian times and continued their evolution through Carboniferous and Permian periods. A number of shark groups became extinct at the end of the Palaeozoic era and they continue today as marine fishes in restricted number. Inspite of the fact that the sharks are the primitive fishes, their cartilagenous skeleton appears to be a secondary development. It is supposed that the bony fishes could have evolved from the primitive cartilagenous fishes by the ossification of skeletal elements. But in recent times, the bony fishes are considered to be the primitive fishes from which the cartilagenous fishes evolved by the reduction of ossification in the skeleton. This view has been supported by the fossil ostracoderms and placoderms which had heavy armour of bones. In addition, the first evolved bony fish is found as fossil in the Silurian period.

The genus *Cladoselache* (Figure 7.4) is the first known shark found in the upper Devonian and it was similar to some of the living forms of today. It had a small shark-like body about three feet in length. The eyes were large and the jaw suspension was of the primitive amphistylic type. The palatoquadrate and Meckel's cartilage were present as single elements provided with primitive teeth. Behind the jaws, there were six pairs of gills. The paired dorsal, pectoral and pelvic fins as well as a pair of small horizontal fins at the base of the tail were present. The caudal fin was of a large heterocercal type. In short, Cladoselache was illustrative of the three important evolutionary contributions of the early fishes namely origin of jaws, origin of teeth and origin of paired fins. Thus this fish had a number of primitive features so that it is considered as the central stem for the later sharks. Then various shark groups such as Pleuracanth sharks, typical sharks, skates and rays as well as bradyodonts and chimaeroids evolved.

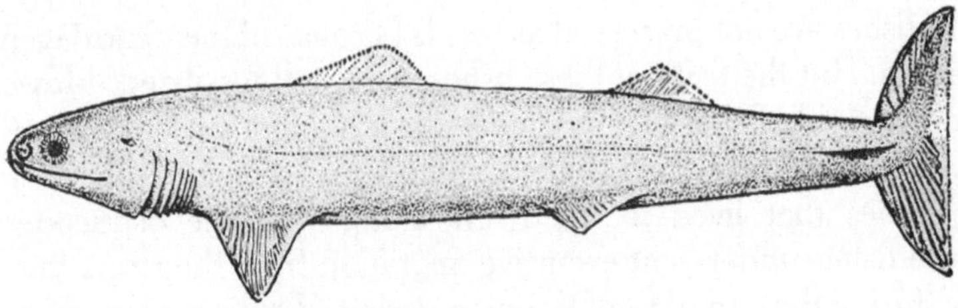

Figure 7.4 *Cladoselache*

The Pleuracanth sharks represented by *Pleuracanthus* and *Xenacanthus* (Figure 7.5) were all fresh water fishes which appeared during Carboniferous and Permian periods. These were more specialized than their Cladoselachian ancestors except having primitive type of amphistylic jaw articulation. They had elongated body with long dorsal fin running along the entire back.

Figure 7.5 *Xenacanthus*

The caudal fin was of a secondarily evolved diphycarecal type. Presence of a long spine directed backwards from the posterior end of the skull was an important characteristic in these sharks.

At the same time, persistant and highly successful groups of sharks evolved in seas all over the world. The primitive earliest representatives of modern sharks were the hypodont sharks. *Hypodus* was the first hypodont shark to appear during the late Devonian and continued through upper Palaeozoic, Mesozoic and even early Cenozoic times. The hypodonts were found to represent intermediate forms during the evolution of cartilagenous fishes. They were primitive in having the teeth derived from Cladoselachians and amphistylic jaw suspension. They were advanced in the respect that they had flexible paired fins with narrow bases. This is the group in which the claspers on the pelvic fins first appeared. Thus these sharks show a close affinity to the modern sharks and an advancement over the Cladoselachians. The hypodonts reached their climax during late Palaeozoic period but they were replaced by the heterodontid and hexanchid sharks as well as a number of other sharks and rays during Mesozoic era. The genus *Heterodontus* (heterodontid) and *Chlamydoselache* and *Hexancus* (hexanchid) form connecting link between the hypodonts and modern sharks. *Heterodontus* can be considered as a slightly modified descendent of the hypodonts.

The modern sharks made their evolutionary history at the advent of the Mesozoic era, especially during Jurassic times. In these fishes, the primitive amphistylic type of jaw suspension was replaced by the hyostylic mode which is the characteristic feature of the more advanced fishes. During this period, higher sharks evolved in two major lines, i.e., one towards typical sharks such as the common sand sharks, the mackeral sharks, the tiger sharks, the white sharks, the basking sharks, the whale sharks and so on and another line towards the highly specialized skates and rays with bottom-living mode of life. Another separate group of shark-like fishes are the Chimaeras (rat-fishes) which are included in the group Holocephali. Only very less number of Chimaeras persist today but a great numbers are known in the form of fossils from early Jurassic times. In addition, the late Palaeozoic sharks namely bradyodonts represented by the genus *Janessa* (ray-like shark) from Permian times were also known. These fishes might have evolved in a number of parallel groups from the primitive ancestral sharks.

Today, most of the shark-like fishes became extinct leaving the group elasmobranchians as living representatives. These are included in two orders namely Selachii (sharks, skates and rays) and Holocephali (chimaeras). Most of these modern sharks have retained more or less the generalized organization of their ancestral forms. The Chimaeras are definitely very ancient fishes and would have evolved from typical sharks during Devonian times.

EVOLUTION OF BONY FISHES (OSTEICHTHYES)

The bony fishes are the most varied and most numerous vertebrates with diversified body forms and aquatic adaptations. These are highly successful group in the animal kingdom and they inhabit both fresh waters and seas, thus attaining the climax of their evolutionary history.

They are highly advanced in most features especially in the ossification of both external and internal skeletons. The bony fishes have well developed bony flap or operculum which cover the entire gill region. In most cases, the jaw suspension is of hyostylic type. Both median and paired fins are supported by bony rays. The lungs are mostly transformed into air bladders that act as hydrostatic structures. Today, the bony fishes represent important vertebrates in the history of evolution and form a dominant group as they introduced the three major traits that all higher vertebrates possess namely legs, skull and bones of skull and limbs.

The earliest known fossil bony fishes are known from the middle Devonian rocks. But the inmediate ancestors of these fishes are not definitely known. It is probable that they might have evolved from shark-like chondrichthyes fishes in the fresh water. Due to competition with the newly evolved bony fishes, the sharks could have migrated to the seas and become exclusively gill breathers. On the other hand, the majority of bony fishes remained in fresh waters and developed lungs or air bladder. Immediately after their origin, the bony fishes became the masters of the water, i.e., the dominant vertebrates of fresh waters.

In the middle Devonian itself, the bony fishes had already split up into three major groups namely Crossopterygii, Dipnoi and Actinopterygii. Of them, the dipnoans (lung fishes) and the crossopterygians (lobe-finned fishes) are air breathers and are included in a subclass namely Choanate or Sarcopterygii in osteichthyes. Phylogenetically, they occupied an important position in the evolutionary history of vertebrates as they played a significant role in the progress beyond the fishes.

The first sarcopterygians such as *Osteolepis* (Crossopterygian) and *Dipterus* (Dipnoan) appeared during Devonian period and were similar to the first actinopterygians like *Cheirolepis* in many respects. In both, the body was fusiform, covered with heavy scales. The skull was shielded by bony plates.

The pectoral fins were close behind the head and the pelvic fins were far back on the body. They had primitive heterocercal tail. But they differed much in many respects as shown in the Table 7.1.

Table 7.1 Differences between ancestral actinopterygians and primitive sacropterygians.

Ancestral actinopterygians	Primitive sacropterygians
Heterocercal tail without epichordal lobe.	Heterocercal tail with a small epichordal love above the body axis.
Fins supported by parallel fin rays.	Archiptery gian type of fins in which the supporting axial elements with lesser bones radiating on sides.
The dorsal fin was single.	There were two dorsal fins.
A pineal opeining present on the top of the head.	The pineal body absent.
Eyes were larger.	Eyes were not so large.
No internal nostrils.	Internal nostrils present.
Scales were of ganoid type.	Scales were of cosmoid type.

The earliest lung fish *Dipterus* (Figure 7.6) had many Choanate features as well as specialized characters such as reduction of bone in the internal skeleton as found in later lung fishes, poorly ossified skull, partially ossified jaws, skull formed by numerous bony plates, teeth adapted for crushing the hard food as in Australian lung fish and so on. They diversified into a number of varieties during late Devonian and late Palaeozoic times. From the dipnoan central line, the lung fish *Ceratodus* arose in Triassic and subsequent periods of Mesozoic era. *Neoceratodus* found in Australia could be a direct descendant of Ceratodus and the lung fishes *Lepidosiren* (South America) and *Protopterus* (Africa) might have evolved as side branches from the central dipnoan stem. It was considered that these fishes gave rise to the amphibians. But as per the opinion of *Romer*, the lung fishes were not the ancestors but 'the uncles of the amphibians'. Now the lung fishes are regarded as somewhat specialized rather than degenerate group because both dipnoans and amphibians are the descendants of a common ancestral stock namely the primitive crossopterygii. It was regarded that the dipnoans would have evolved as a side branch of the bony fishes from the primitive crossopterygians during middle Devonian. They were found plenty during the Palaeozoic time and gradually declined with a few surviving representatives (*Protopterus*,

Lepidosiren and *Neoceratodus* as shown in Figure 7.7). These lung fishes had worldwide distribution in the past but they are very much restricted to the three regions(Australia, South America and Africa respectively) in recent times. Thus they illustrate the best example for the discontinuous distribution of species.

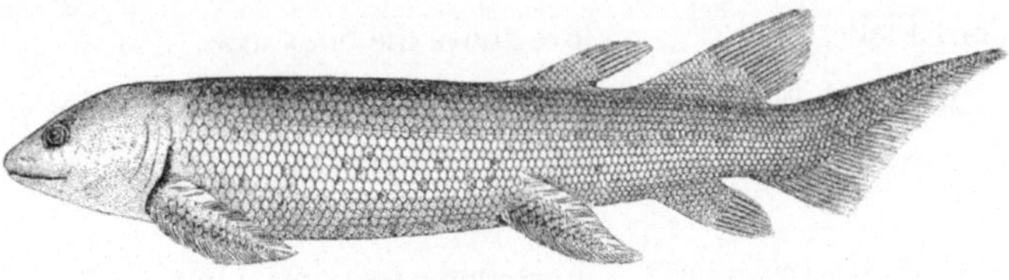

Figure 7.6 *Dipterus*

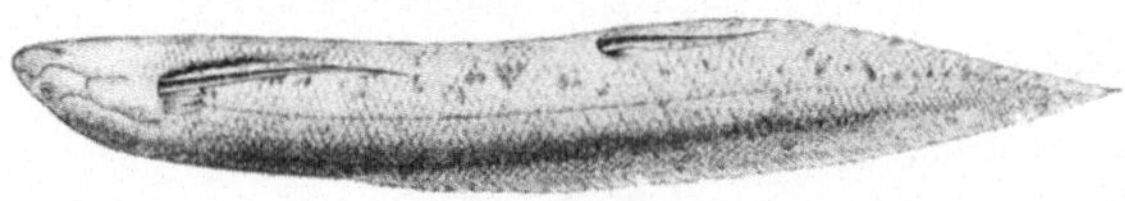

(a) Protopterus

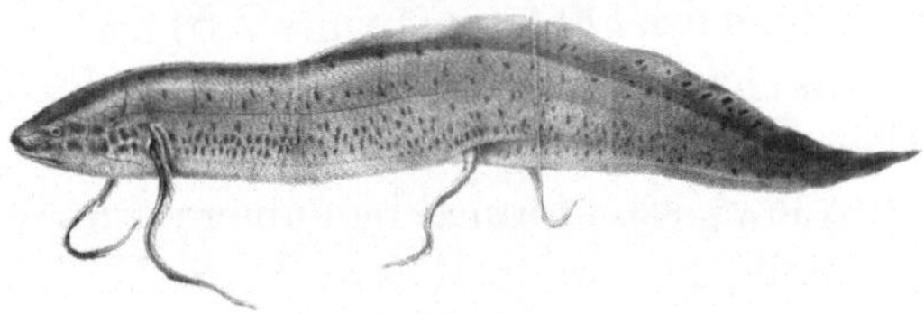

(b) Lepidosiren

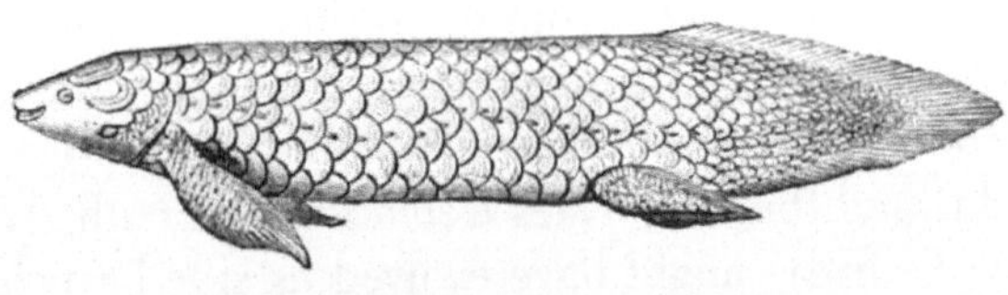

(c) Neoceratodus

Figure 7.7 Lung Fishes

The crossopterygians which were the lobe-finned fishes abundant during Devonian times, declined during Carboniferous period and became almost extinct before the end of Palaeozoic era. A few Crossopterygians survived during Mesozoic era and one of its branches gave rise to the primitive amphibians. The Crossopterygian fishes also appeared in the middle Devonian and they had general choanate characters.

Osteolepis (Figure 7.8) was an early Devonian crossopterygian.It had fusiform body covered with cosmoid scales. The paired fins were lobate and archipterygial type along with two dorsal fins and a heterocercal caudal fin. It had a strong notochord. The skull and the jaws were completely bony and were comparable to the pattern of other bony fishes and early amphibians, thus indicating that they were towards the direct line between fishes and amphibians. Osteolepis and other Devonian crossopterygians had highly ossified skull comprising a number of bones. They were carnivorous in habit and possessed sharp pointed labyrinthodont teeth as also seen in the early amphibians. These fishes had well developed internal nostrils. In early crossopterygians, the fins especially the pectoral fins had single proximal bone which articulated with the girdle. Below this, there were two pieces of bones as well as radiating bones towards the distal edges of the fins. This type of bone arrangements in the fins could be the starting point for the evolution of the limb bones in land-living vertebrates. Therefore the crossopterygians are considered to be the direct ancestors of the first amphibians.

Figure 7.8 *Osteolepis*

From the Devonian basal stock, the crossopterygians evolved into two major groups namely Rhipidistia and Coelocanthini. The osteolepids were included in Rhipidistia which further diverged into two lines, one leading to holoptychians in late Devonian as a side branch of the lung fishes and the another one towards the rhizodonts represented by the Devonian genus *Eusthenopteron* (Figure 7.9). The rhizodont fishes were more advanced than their osteolepid ancestors. Eusthenopteron had an elongated body with a skull pattern resembling the skull of early amphibians. It had advanced vertebrae, symmetrical tail and the vertebral column extending back to the tip. The paired fins were seen to progress towards the tetrapod limb. The *coelacanthus* were marine fishes with deep body, lobate paired fins having reduced internal elements, a diphycercal tail bearing an additional small lobe and calcified lungs.

The typical members of this group included *coelacanthus* from lower Carboniferous and Permian periods and Cretaceous *Macropoma*.

Figure 7.9 *Eusthenopteron*

It is believed that the coelacanths became extinct at the end of cretaceous but a living coelacanth was first found in South Africa in 1938. It was named as *Latimeria* (Figure 7.10) which is the only living representative of the group crossopterygii. So, it is considered as a 'living fossil'. It closely resembles Macropoma and has a large body covered by brilliant scales, a single spiracle, long and strong lobed paired fins along with two dorsal and one diphycercal tail fin and a pair of rostral organs. A number of specimens of Latimeria have been known from time to time near Madagascar and in the Comoro islands between Madagascar and Africa.

Figure 7.10 *Latimeria*

The actinopterygians were the ray-finned fishes which appeared in the middle Devonian times. The earliest known ray-finned fishes were Palaeoniscoids which would have derived from lobe- finned fishes. *Cheirolepis* (Figure 7.11), *Palaeoniscus*, *Amphicentrum* and *Dorypterus* were the Palaeoniscoid fossil fishes found in the middle Devonian deposits. These were regarded as the first formed early bony fishes. *Cheirolepis* was a small fish with a fusiform body covered by rhomboid scales. It had ossified skull, single dorsal fin, paired pectoral and pelvic fins and a heterocercal caudal fin.

Figure 7.11 *Cheirolepis*

The vertebrae were incompletely ossified but the notochord was well developed. *Palaeoniscus* was also a primitive actinopterygian in Permian times and was more advanced than cheirolepis in few respects. *Amphicentrum* was a deep-bodied, primitive actinopterygian from Pennsylvanian and Permian periods with elongated dorsal fin and heterocercal caudal fin. In *Dorypterus*, the anterior part of the dorsal fin was enormously elongated and the pelvic fins were seen to migrate forward towards the throat in front of the pectoral fins as in higher teleost fishes. These palaeoniscoids evolved through a few general stages with a handful of surviving representatives which are collectively designated as the ganoid fishes. These are included in groups namely Polypterini, chondrostei, Holostei and Teleostei.

The polypterinean fishes are regarded as extremely isolated ray-finned fishes without having close resemblance among the extinct and living forms. Most probably, these might be the nearest relatives of the palaeoniscoids and they comprise two surviving genera namely *Polypterus* and *Calamoichthyes*. The chondrosteans evolved during late Palaeozoic and were characterized by having reduced scales in the body. In the genus *Tarrasius*, there is not only loss of scales but also the pelvic fins. Moreover, the median fins are fused to form a continuous fin on the back around the tail and along the ventral side. Thus these fishes show primitive, specialized and degenerate characteristics and include five surviving genera namely *Acipencer*, *Scaphyrhynchus*, *Polyodon*, *Crossopholis* and spoon bills. The fossil fishes known as *Redfieldia* of Triassic times are considered to be advanced palaeoniscoids as well as advanced chondrosteans progressing towards holosteans. In other words, they were the intermediates between chondrosteans and holosteans.

The holosteans arose from several basic stems. These were more specialized than the chondrosteans especially in the skull, jaws and vertebrate along with reduction in the fin rays and scale structure and loss of spiracles. But the holosteans became extinct after the early Cretaceous period except two surviving archaic groups namely Amioidei and Lepidosteoidei. All the amioideans became extinct except one species namely *Amia* which resemble all surviving primitive bony fishes in structure and form. It is believed that amioideans might have given rise to the teleostean fishes. The group *Lepidosteoidei* is represented by a single genus *Lepidosteus* which is relatively modern but possesses many primitive characters.

The teleosteans which include the familiar examples such as salmon, cat fishes, perches, eels, *Notopterus*, *Catla*, hippocampus, mugil, tetrodon, *Anabas*, sucker fish, etc., are the most successful survivors of the ancestral ray-finned fishes. They might have arisen from the holostean stock most probably from Amioidea during Jurassic period. A Jurassic fossil *Leptolepis* could be the first teleostean fishes originated from a primitive holostean stock during Jurassic times. The teleosteans have multiplied very rapidly and underwent extensive adaptive radiation during Cenozoic times. However, they have not led to any higher levels of evolution except the specialized types of fishes.

From the foregoing account, it is clear that the ostracoderms are the first known vertebrate fossils. They might be bottom detritus feeders as they had no jaws but with hard bony armour. The present day cyclostomes are the living representatives of this group in which the bones have degenerated so that they possess a cartilageneous skeleton. The placoderms are the first jawed vertebrates which might have arisen from an ostracoderm-like ancestor by the conversion of the first gill arch into upper and lower jaws.

Instead of the bottom detritus feeding they began to feed on a variety of prey and diversified into fish-like heterogenous vertebrate groups. Some of the placoderms like acanthodians had both pectoral and pelvic fins in between the two series of paired finlets, thus indicating the origin of fins from the original continuous fin fold.

Acanthodian-like ancestors gave rise to both cartilagenous and bony fishes. The Cartilagenous fishes evolved by the degeneration of bone and they soon diversified into typical sharks, skates, rays and holocephalians. These are highly successful group and continue to be so at present day.

It is evident that even at the beginning, the bony fishes were of two distinct groups namely lobe-finned fishes (crossopterygians) and ray-finned fishes (actinopterygians). The lung fishes are related to the lobe-finned fishes. They were numerous in number and became highly specialized and are represented today by *Neoceratodus, Protoptenus* and *Lepidosiren*.

The stem group of ray-finned fishes was palaeoniscoids which gave rise to a variety of fish groups like chondrostei, Holostei and Teleostei. The present day *Acipencer* represents the chondrostean level and *Amia* exhibits the holostean level of evolution. Teleostei flourished from Jurassic times and are the climax group today occuring in enormous numbers adapted for different types of aquatic life.

The lobe-finned fishes were similar to the lung fishes so that both are included in one group namely choanichthyes. They flourished during Devonian times and became extinct except for a living species *Latimeria*. The crossopterygians are of prime importance in the respect that they gave rise to the land vertebrates. The Devonian period was a time of seasonal droughts so that these fishes had to migrate fromone pool to another in search of water. In this effort, they developed lungs to struggle over the land. In addition, they walked almost by the fins with the basal lobes.

The fin skeletons of these fishes became the basis for the pentadactyl limb of the land vertebrates. Those fishes which were capable of living longer on the land, had more chances of survival and became the first amphibians about 300 million years ago. Thus the origin of first amphibians from the lobe-finned fishes was an easy transition.

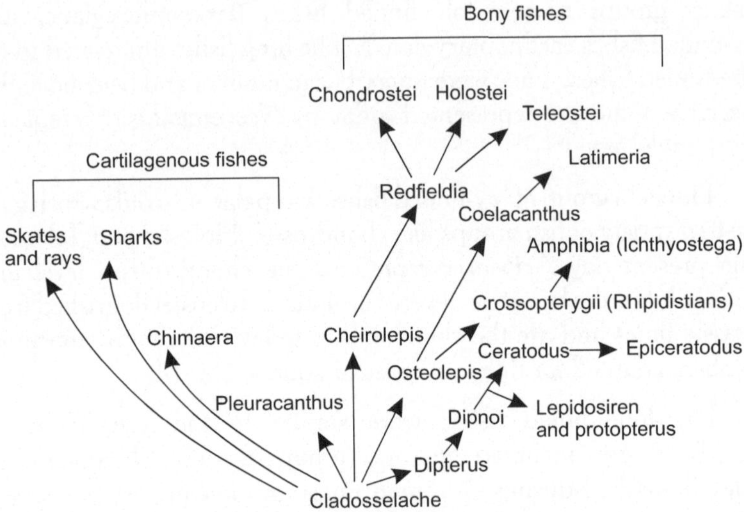

Evolutionary tree of cartilagenous and bony fishes

QUESTIONS

1. Write an account of the three major evolutionary events during the evolution of fishes.
2. Discuss various theories regarding the origin of paired fins in fishes.
3. Explain the origin and evolution of cartilagenous / bony fishes.
4. Write an essay on extinct fossils of cartilagenous / bony fishes.
5. Describe the evolutionary significance of crossopterygians.
6. Briefly explain the evolutionary tree of cartilagenous / bony fishes.
7. Write notes on the following :
   a) *Cladoselache* b) Hypodont sharks
   c) *Chimaera* d) Sarcopterygians
   e) Lobe-finned fishes /*Neoceratodus Lepidosiren/Protopterus*
   f) *Osteolepis* g) *Eusthenopteron*
   h) *Latimeria* i) Palaeoniscoids

8

EVOLUTION OF AMPHIBIANS

The amphibians include a group of cold-blooded vertebrates that lie in an intermediate position between the evolutionary history of fishes and reptiles. These are the first vertebrates to move from an aquatic habitat to a terrestrial one and are the ancestors of all reptiles, birds and mammals. They emerged in late Devonian and diversified extensively in the Carboniferous, Permian and Triassic times reaching the climax. By the end of Triassic, all the ancient groups of amphibians became extinct due to their inability to compete with the advanced amphibians and reptiles. A second radiation of amphibians occured in the Jurassic period and has continued to the present day. Varied groups of vertebrates made many attempts to invade the land to live in the terrestrial environment. The Devonian period was a time of seasonal droughts during which the crossopterygian fishes could have crawled across the land to reach the water. From such a start, the definite terrestrial amphibians developed during the drought to cross over the land. They might have come out of the water to find new ponds as well as for the available food sources on land. It is a widely accepted fact that the amphibians have originated from the fishes. There are many similarities between the fishes and the living amphibians and so Huxley grouped both the groups in a single class Ichthyostega. It seems that the fishes themselves acquired many of the changes as adaptations to the seasonal droughts even before the amphibians were actually evolved.

ORIGIN AND EVOLUTION

Fossil studies indicate that the amphibians arose at the time when the crossopterygian fishes were abundant (Devonian period). A rhipidistian crossopterygian fish namely Eusthenopteron would be the direct ancestor which became transformed into the first land vertebrates, at the end of Devonian during the transition from the Devonian to Mississippian times. The crostopperygian fishes had many characteristic features on line with the amphibians. The osteolepid crossopterygians fishes were the nearest relatives of amphibians. *Panderichthyes* was a lobe-finned osteolepid fish with the skull bone similar to the earliest tetrapods. The fin skeleton of these fishes resembled the fore and hind limbs of tetrapods. Its brain case and teeth were similar to the tetrapods. At the same time, it had lungs, nostrils and gills as well as a hole between the nasal passage and the mouth as in other osteolepid fishes. It had a tail resembling amphibians but had no dorsal and anal fins. *Sauripterus* was on direct line of descent between fishes and amphibians. It had fins with eight fingers similar to the early amphibians. *Densignathus roweri* was a transitional form with jaws resembling fishes and stem tetrapods.

Ichthyostegids represented by Acanthostega, Ichthyostega and Ichthyostegopsis found in the upper Devonian period were the intermediates between the osteolepids and early amphibians. Acanthostega was clearly a tetrapod but it never left the water. Its legs were not suited to support the body weight on land. The wrists were absent but the limbs had well developed digits (8 in fore limbs and 7 in hind limbs). It had more fish-like forelegs, gills and caudal fin. Ichthyostega (Figure 8.1) was more terrestrial and their legs were aquatic and was well on the way to the full pledged amphibians. It was the first animal to have the feet with more digits than tetropods. The skull was solid and completely roofed as in crossopterygian fishes but the post cranial skeleton was a strange mixture of fishes and amphibians. The vertebrae progressed little beyond the fishes. It had external nasal openings but the internal gills were absent. The feet had more digits than in tetrapods but with fewer bony rays than the fishes. While these feet were unable to support the weight of the body, they were good for being in water. The fore limbs were more robust but the hind limbs were more like paddles with seven digits. In general, these earliest tetrapods had only partially evolved limbs as the hands were able to bend slightly. They had well developed gills and were found in river sediments suggesting their fresh water habitats. They were still aquatic as they retained the gills. Pederpes finneyae was a transitional form between the Ichthyostegid tetrapods and later amphibians. It possessed intermediate traits in ear bones, ribs, metacarpals, etc. and had five toes in the feet.

Figure 8.1 *Ichthyostega*

The following Table 8.1 gives the major differences between the crossopterygian fishes and primitive amphibians.

TABLE 8.1 Differences between the Crossopterygians and Ichthyostegids

Crossopterygians	Ichthyostegids
Lungs were present but gills were primarily used.	Lungs were the primary respiratory organs.
No specialization of skin to prevent desiccation.	Skin was specialized to prevent desiccation.
Skeleton was not stronger.	Stronger skeleton with increased gravity on land.
Body and tail were used for locomotion and the fins for balance.	Hind limbs were used for locomotion and tail for balance

The main early tetrapod evolution took place in the Carboniferous period about 350–290 million years ago. The newly formed habitats encouraged a number of possibilities to the early tetrapods for extensive diversification. In the fossil record, the early Carboniferous tetrapods are poorly known whereas the late Carboniferous tetrapods are well known. One of the basal tetrapods were labyrinthodont amphibians (Labyrinthodontia or Stegocephalia) which appeared at the end of the Devonian period. Among the labyrinthodont amphibians, the anthracosaurs were the early descendants of Ichthyostegids during Mississippian. These amphibians evolved in two lines namely Embolomeres and Seymouriamorphs.

The embolomeres had anthracosaurian type of vertebrae which were well adapted for supporting the body on land. In these vertebrae, an intercentrum and pleurocentrum were present one behind the other on which the neural arch and the spine were placed. *Pteroplax* or *Eogyrinus* was a typical embolomerian genus and *Pholidogaster* and *Diplovertebron* were the oldest anthracosaurs and were more primitive than the embolomeres.

All these embolomeres had characteristics similar to the primitive amphibians. The seymouriamorphs arose from common ancestors with other anthracosaurian amphibians. *Seymouria* was a small tetrapod from lower Permian and was almost the intermediate form between the amphibians and reptiles. It had a completely roofed labyrinthodont skull, teeth and single occipital condyle. The postcranial skeleton was progressive towards the early reptiles.

Discosauriscus was closely related to Seymouria. Though seymouria possesses a mixture of amphibian and reptilian characters, it is now considered to be an anthracosaurian labyrinthodont. Moreover, these amphibians were not the reptilian ancestors but they evolved in the direction of reptiles.

Diadectomorphs (*Diadectes*) were very close to the seymouriamorphs from the lower Permian beds. Diadectes and its relatives possessed many primitive characters of seymouriamorphs as well as many reptilian characters along with a trend towards large size.

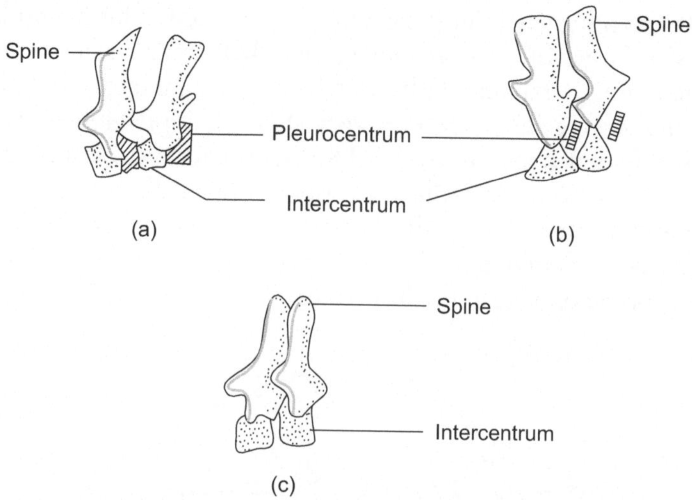

Figure 8.2 Types of vertebrae in different amphibian groups

The temnospondyles were an important group of labyrinthodont amphibians which evolved in the early amphibian history. The earliest temnospondyles were known as rachitomes which originated from the ancestral ichthyostegids during Mississippian period. They had rachitomous type of vertebrae in which the intercentrum was large and wedge-shaped and the Pleurocentrum was small, fitted in between the intercentra and both supporting the neural arch of each vertebra (Figure 8.2). Eryops (Figure 8.3) from the lower Permian period was the typical genus of this group.

Figure 8.3 Eryops

It was a heavily built large animal well adapted for life on land. It had a large, broad, flat and roofed skull, very strong vertebral column, greatly expanded ribs, and heavy pectoral girdle and pelvis. The limbs were short but were very stout. The rachitomes evolved along different lines during the Pennsylvanian and Permian times with varying adaptive radiation. Among them, *Trimerorachis* from the Permian was a small rhachitome with a foot or two in length. It had a much flattened skull with the body covered by the overlapping scales corresponding to the fish scales.

Figure 8.4 Cacopus

It was mainly a water-living amphibian found in shallow streams and ponds. *Cacopus* (Figure 8.4) was another small rhachitome from the Permian times with a heavy skull, short body, strong legs and reduced tail. In these animals, the body was covered with heavy bony armour at the back. *Branchiosaurs* (Figure 8.5) were very small labyrinthodonts found as skeletons from late Pennsylvanian and early Permian times.

Figure 8.5 Branchiosaurus

They were included in a separate group namely Phyllospondyli by some Palaeontologists. But Romer proved that these would probably be larval forms of some rhachitomes. The rhachitomes were the major dominant group during Permian times but they began to decline at the end of Permian period. Trematosaurs were a few rhachitomes existing to the early Triassic. They had advanced skull towards the stereospondyls but their vertebrae were rachitomous type. These were marine and fish-eating animals and they disappeared soon after giving rise to the stereospondyls.

The stereospondyls were highly specialized aquatic tetrapods in Triassic times. They had secondarily simplified vertebral column in which the vertebral centra were reduced with the loss of pleurocentra (stereospondylous type of vertebrae). Therefore, there was a general trend in these amphibians towards an increase in size of the body. The last labyrinthodonts were represented by Metoposaurus which was a large stereospondyl amphibian. It had a very large skull when compared with the flattened body. The skeleton had increased cartilagenous elements with the reduction of bony materials. In short, these animals exhibited a strange combination of characters. Plagiosaurs were the aberrant group of stereospondyls in the late Triassic times. They had an extremely flat and broad skull and body but the limbs were small. *Gerrothorax* of this group had external gills as in modern salamanders. At the end of Triassic, the stereospondyls and Plagiosaurs became extinct along with a number of land-living vertebrate animals, thus making an end to the history of the labyrinthodonts.

The lepospondyls were numerous in number and were dominant during the late Palaeozoic and Triassic periods, sharing the fresh waters and land with labyrinthodonts. These comprised highly varied groups that appeared during early Mississippian. They were dominant during the Pennsylvanian and Permian times and became extinct at the end of Palaeozoic. Many lepospondyls were small, primitive and adapted to live at the edge of waters and in the swamps.

Microbrachis (*Microsauria*) and snake-like elongated *Ophiderpeton* (*Aistopoda*) of Pennsylvanian were the representatives of this group. Another group of Palaeozoic lepospondyl amphibians were Nectridians and they evolved into two lines during the Pennsylvanian period. In one group, the body became elongated and snake-like as seen in aistopods and in another line, the body and skull became flatened and broadened. The genus *Sauropleura* had a pointed skull and long body but the limbs were absent.

The evolutionary development towards the flattening of the skull and body was exemplified by *Diplocaulus* (Figure 8.6) in which the body was flat and the skull had lateral projections but the legs were very small and weak.

Figure 8.6 *Diplocaulus*

The three recent amphibian groups namely Anura, Urodela (Caudata) and Apoda are regarded as Lissamphibia by the modern biologists. All these share certain common characteristics, especially in the teeth which are bicuspid and pedicellate. In general, the ancestry of lissamphibians can be traced from the Temnospondyls. A temnospondyl genus *Dolesserperton* had pedicellate teeth and some like *Branchiosaurus* possessed certain lissamphibian features. The fossil genus *Gerobatrachus hottoni* ('Hotton's elder frog') discovered from Texas in 1995 represents exactly the intermediate condition between the more primitive amphibians and both the frogs and salamanders. Its wide and highly built skull, backbone and teeth were similar to the frogs and Salamanders. These amphibians might have evolved in two distinct lines between 275–240 million years ago. Small labyrinthodonts like dissorophids had specialization in the ear region with large tympanum supported by the squamosal and small stapes. Hence, the labyrinthodonts are also believed to be the likely ancestors of frogs. The genus *Triadobatrachus* of lower Triassic had moderately short body with a remnant tail. The skull bones were greatly rduced in number, thus approaching a condition towards modern frogs and toads. The ribs were well developed and the pelvic girdle had an elongated ilium. The limb bones were similar to the frogs. Therefore, it is assumed that the modern anurans would have evolved from Triradobatrachus at the beginning of Jurassic period by a sudden and complete change in the skeleton.

The basic structure of the skull and its roof, especially the area of the middle ear of Microsaurs (lepospondyls) closely resemble the apodans and salamanders, showing the possibility of the origin of these amphibians from lepospondyls. A most primitive caudate namely *Hynobius*, a land salamander was very close to the central stock. In salamanders, the bones are secondarily lost and the loss is substituted by the cartilagenous skeleton.

Thus the amphibians were the first vertebrates to conquer the land as an offshoot from one group of Crossopterygii namely Osteolepidoti. But they were not destined to compete the conquest because 'they are still chained to water; in water they are born; to water they must peiodically return for reproduction'. The Palaeozoic extinct amphibians have left some surviving groups. Among them, Apoda or Gymnophiona is regarded as the primitive modern amphibian group specilized for borrowing mode of life with many degenerate characters. The caudates or Urodeles form a natural group comprising Newts and Salamanders. The frogs and toads which are included in Salientia are at the climax forming a natural and well defined group occuring in a great variety of habitats.

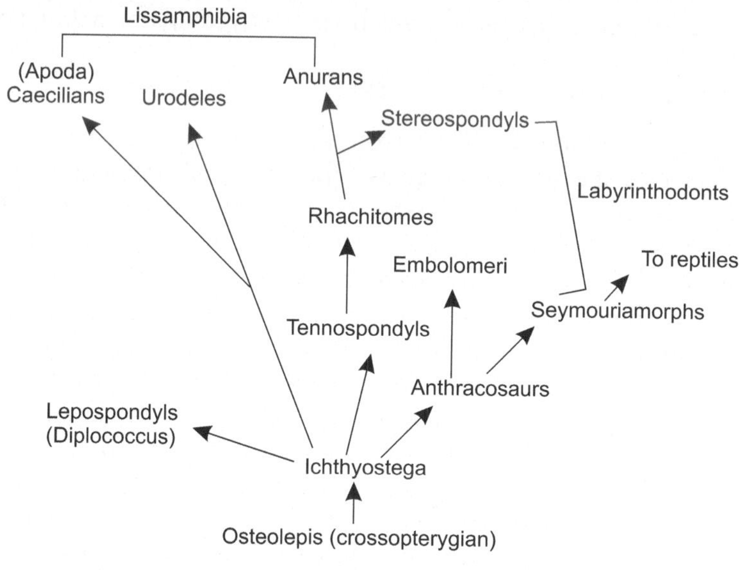

Evolutionary tree of amphibians

CHANGES DURING THE TRANSITIONAL PERIOD FROM AQUATIC LIFE TO LAND LIFE IN AMPHIBIAN EVOLUTION (TERRESTRIALIZATION)

A significant evolutionary trend has been noticed when certain vertebrates (fishes) left the water and stepped on to the land as the first amphibians in Devonian times. For the first time, the vertebrates were invading a new environment that was very different from the one in which they had lived for millions of years. Thus the amphibian evolution was a remarkable step that involved great many changes in almost all systems of the body of the vertebrate structure during the transitional period from the aquatic life to the land life. The adaptations acqiured by the first land vertebrates to live on land is referred to as terrestrialization which seemed to be started even in early amphibians. The ichthyostegids had all the basic reqiurements for life on land at the end of Devonian. During Mississippian and Pennsylvanian times, the primary characteristics of the tetrapods were established in the successors of ichthyostegids namely the labyrinthodont amphibians or stegocephalians. The following account gives the important changes that occured in the structure of vertebrates during amphibian evolution.

Changes in the Respiratory System

In early amphibians, the terrestrial respiration (air-breathing) was one of the major problems. But this problem had already been solved in the ancestors of amphibians namely the Crossopterygians which had developed lungs. However, in most of these fishes, gills were the primary respiratory organs and the lungs were used as accessory respiratory structures. This situation was reversed in most of the amphibians. The fossil evidences indicate that the first land vertebrates were primarily air breathers and the gills were the respiratory organs in the larval stages as in the case of the present day amphibians. With the perfection of lungs, the branchial structures assumed several other functions. The spiracle was modified into the tympanic cavity of the ear and the skin covering the spiracle became the ear drum. The hyomandibular piece became modified into the columella auris to connect the tympanum and inner ear. The hyoid apparatus which developed from the branchial arches supported the tongue.

Changes in the Body Covering

The another important difficult situation faced by the first land vertebrates was desiccation, i.e., they had to retain their body fluid when they were out of water. To achieve this, the early amphibians could not have travelled very far away places from the water sources and they would have returned to water frequently. Moreover, these animals might have developed body coverings against desiccation. Evidences show that the body of first amphibians were covered by scales and tough skin provided with ossicles in order to prevent the evaporation of body fluid.

Changes to Manage the Gravitational Force

The gravity is a strong factor that influences the structure and life of animals on land. When the fish is in the water, its body weight is effectively zero. But on land, the body has to be held up by the limbs. Therefore, the skeleton and all the internal organs have to be structurally modified to cope up with the gravitational force. To overcome the gravity, the first land vertebrates developed strong backbone, muscles and limbs. In early amphibians, the land locomotion was aided by the limbs with well developed pectoral and pelvic girdles to hold the body against gravity and to propel the animal with tail which was used as a balancing organ. In addition, the connection between the pectoral girdle and the skull was lost and the pelvic girdle became attached to the posterior part of the vertebral column. Again, the pectoral girdle was strengthened by the union of two halves to bear the entire weight of the body on land.

Changes in Circulatory System

As a consequence of the disappearance of gills in amphibians, the circulatory system became modified to suit the pulmonary respiration. The heart became three chambered with two auricles and one ventricle and adapted to pump venous blood to the lungs for purification. Of the four pairs of arteries of the piscean ventral aorta, the first pair were modified into carotid arteries, the second pair into systemic artery which unite to form dorsal aorta and fourth arches changed into the pulmocutaneous arteries with the disappearance of third aortic arches. The iliac artery was transformed into femoral and sciatic arteries and the post caval appeared to take the blood to the heart from the kidneys.

Origin of Pentadactyl Limb

Of all the changes that occured during the evolution of the first land vertebrates, the origin of pentadactyl limb was a major change for the locomotion on land. The paired fins of fishes have no value on land so that it is important to be modified into efficient paired land limbs. Even though the eusthenopteran fishes possessed skeleton of pectoral and pelvic fins like those of early tetrapods, they could have not walked properly with its fin on land. Moreover, the limbs were pointed backwards and sideways in these fishes. Therefore, for the ability to walk on land, considerable modifications such as appearance of new bones, well-defined elbow and wrist joints, lengthening of humerus, etc., might have occurred in the tetrapod limbs.

In osteolepid fishes, the base of the fin became narrow and the radials were concentrated at the base and free at the tip in the form of a fan with the central radial forming the central axis. This central axis might have elongated in the fins of dipnoans and this type of limb is known as archipterygium. In the case of higher bony fishes, the central axis is reduced in the fins and this type is known as ichthyopterygium (Figure 8.7).

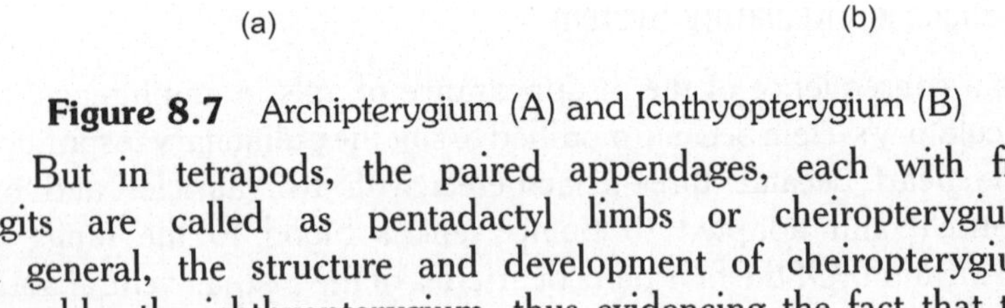

Figure 8.7 Archipterygium (A) and Ichthyopterygium (B)

But in tetrapods, the paired appendages, each with five digits are called as pentadactyl limbs or cheiropterygium. In general, the structure and development of cheiropterygium resembles the ichthyopterygium, thus evidencing the fact that the lobed fins of crossopterygians were the basis for the appearance of the pentadactyl limbs. The tetrapod limb can be easily derived from the ichthyopterygium of crossopterygian fishes by 1) the elongation of the bones of the basal lobe, 2) the loss of expanded fin blades and 3) the modification of the radials into digits as shown in the Figure 8.8.

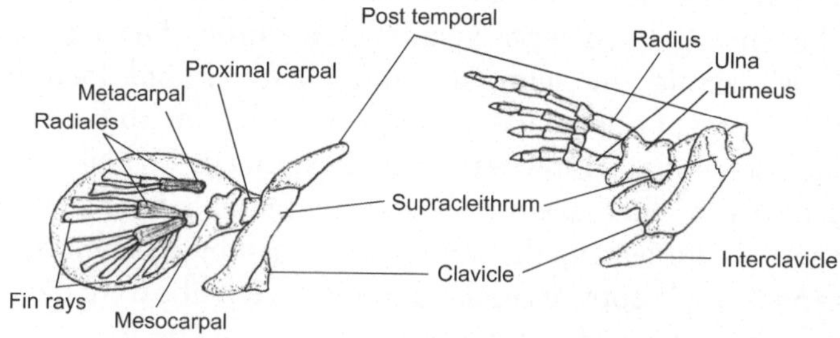

Figure 8.8 Lobed fin of Sauripterus (A) and Pentadactyl limb of Eryops (B)

There are many theories to explain the origin of pentadactyl limb from the paired fins. According to the axial theory, the cheriopterygium could have evolved from a structure which was very powerful to bear the weight of the body on land and so the fin might be transformed into the pentadactyl limb by theloss of fin rays and reduction of distal elements. According to the Basal theory, the fleshy base of the fin of the eusthenopteron fish (Figure 8.9) had a small propterygium, a large mesopterygium and a posterior metapterygium of which the mesopterygium gave rise to the humerus in the forelimb or femur in the hind limb, the propterygium to the radius or tibia, the metapterygium to the ulna or fibula and the radials into a series of bones in the digits. There is no doubt that the pectoral fins of crossopterygians contained major proximal bones of a tetrapod limb namely a single humerus, radius and ulna as also seen in early tetrapods.

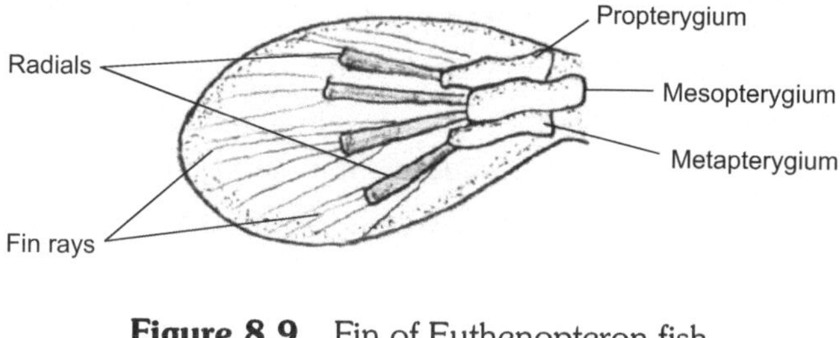

Figure 8.9 Fin of Euthenopteron fish

The above account clearly envisages that the ancestors of tetrapods might have possessed similar pectoral and pelvic fins. In addition, most of the primitive amphibians could have lived along with the ancestral crossopterygian fishes during Devonian times. Both were similar in many structural features and food habits. The development of pentadactyl limb could be due to the necessity for land locomotion. However, the development of limbs and the ability to live on land seem to be the adaptations for remaining in water and not for leaving the water. As opined by Romer, the origin of tetrapods is a 'happy accident' and the land limbs were developed 'to reach the water and not to leave it'.

In addition to the above major and essential adaptations for the terrestrial mode of life, the present amphibians possess smooth skin without scales except caecilians. The skull has two occipital condyles to fit with the atlas. The lateral line system of fishes is completely reduced in amphibians. The pulmonary respiration is supplemented by the development of cutaneous respiration by the skin. The life history of modern amphibians include fish-like larval stages. Though the amphibians developed many adaptations to live on land during its evolutionary course, they never solved the problem of reproduction away from water. Therefore, the amphibians have been forced to return to water for egg laying. As quoted by Romer, the amphibians are still chained to water as they are unable to conquer the land completely.

QUESTIONS

1. Trace the origin and evolution of amphibians.
2. Give an account of amphibian fossils.
3. Explain the evolutionary significance of labyrinthodonts.
4. Write an essay on the importance of ichthyostegids.
5. Describe the trends in the evolution of recent amphibians.
6. What are the important changes that took place during the transitional period from aquatic life to land life?
7. What is terrestrialzation? Explain this process in detail.
8. Give an account on the origin of pentadactyl limb.

9. Write notes on :
   a) Ichthyostega
   b) Labyrinthodont amphibians
   c) *Seymouria*
   d) *Eryops*
   e) Lissamphibia
   f) Terrestrialization
   g) Cheiropterygium

9

EVOLUTION OF REPTILES

The reptiles are the first true terrestrial vertebrates with amniotic eggs. They first evolved from amphibians about 200 million years ago in Carboniferous period and were dominant in world's fauna during Mesozoic era. The reptilian fossils that possess intermediate characteristics between the amphibians and reptiles provide ample evidence for the evolution of reptiles from amphibians. While the temnospondyl and anthracosaur amphibians were prevalent on land during late Carboniferous times, small lizard-like tetrapods which laid land eggs were also found. These were the first amniotes which were the ancestors of all the major tetrapod groups dominating the land from Permian onwards. They have many advanced features over the amphibians such as internal fertilization with shelled egg (amniotic egg), more independency from water, thick impermeable skin, strong limbs and so on. They laid fewer eggs than the amphibians and fishes did. The evolution of reptiles happened to be over a period of millions of years. They evolved into different groups with varied characteristics for the survival in diverged environments. Thus the evolution of reptiles from the ancestral stock was on different lines resulting in all the living and fossil forms and so it provides one of the best examples for macroevolution or adaptive radiation. They occupy an evolutionary position between the amphibians in one hand and the birds and mammals on the other hand.

The appearance of amniotic egg or land egg was a great forward step in the evolution of vertebrates during Carboniferous times. Because, with this type of egg, the reptiles would have separated and wandered freely on the land without returning to the water for reproduction. But the fossil evidences of amniotic eggs or embryos of primitive reptiles are not available. The oldest known fossil amniotic egg belongs to the lower Permian which was a period long after the well establishment of the reptiles. The present day small Plethodont salamanders lay their eggs in damp places on land rather than in water. Therefore, it is probable that the first amniotic eggs might have evolved from such amphibious terrestrial eggs.

Itisthefactthatthereptilesarosefromaprimitivegroupofamphibians namely the labyrinthodonts, especially from the anthracosaurians. The first reptiles that originated from the primitive amphibian stock during late Pennsylvanian and early Permian times are designated as stem reptiles which are included in the group Cotlyosauria. This basic cotlyosaurian stem gave rise to all higher reptiles between the Permian and Cretaceous times. They acquired numerous adaptations to lead a successful life on land and conquered the land with the origin of land eggs and scales. In the labyrinthodont ancestors and the cotlyosaurs, the skull roof was solid with openings for the nostrils, eyes and pineal body. As the reptiles proceeded to evolve, the skull roof developed additional openings behind the orbit (temporal fenestrae) to give accommodation for the jaw muscles and to reduce the weight of the skull. Thus four amniotic skull types evolved (Figure 9.1). The anapsid type of skull had no fossae as seen in stem reptiles, chelonians and mesosaurians. The diapsid type had two fossae as in lizards, snakes, crocodiles and birds as well as in extinct dinosaurs and pterosaurs. The synapsid type had only one lower fossa as in extinct pelycosaurs and synapsids.

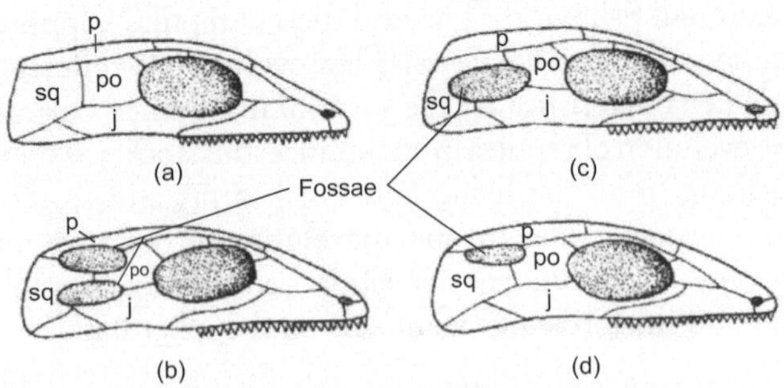

Figure 9.1 Amniotic skull types

The Parapsid (Euryapsida) type had one upper fossa as in later amniotes such as Nothosaurs, Plesiosaurs and Ichthyosaurs of Mesozoic times. In the mid-Carboniferous, the amniotic evolution resulted in the diversion of two great groups namely Synapsida leading to mammals and Sauropsida (Anapsida and Diapsida) leading to chelorians, lizards, snakes, crocodiles and birds.

ANAPSID REPTILES

The most known primitive fossil stem reptile was *Seymouria* from the Permian sediments. The other completely preserved cotylosaurian reptiles were *Limnoscelis*, *Hylonomous*, Captorhinus, *Procolophon*, *Diadectes*, etc. These are considered to be the most primitive reptiles that resembled the labyrinthodont ancestors. The evolution of the first Cotylosaurs from the labyrinthodonts was a major step in the vertebrate history and in the subsequent reptilian history.Seymouria was an intermediate form between the amphibians (Embolomeri) and reptiles (Cotylosauria) and it gave rise to all other reptiles. As it possessed both amphibian and reptilian features, it forms a connecting link or intermediate form between the two groups. According to Colbert, it is a 'basic primitive reptile'. But Romer places seymouria in the amphibian order Anthracosauria as it possesses more amphibian characters than the reptilian features.

Seymouria (Figure 9.2) was a lizard-like terrestrial carnivore with pointed head and short tail. In the following respects, it resembled the amphibians and reptiles.

Figure 9.2 *Seymouria*

Amphibian characters

1. Nature and arrangement of skull bones are similar to that of Embolomeri.
2. Presence of two occipital condyles, labyrinthodont teeth, otic or auditory notch in the skull and lateral line system.

Reptilian characters

1. The post cranial skeleton is similar to the cotylosaurs.
2. Presence of reptilian vertebrae. In the vertebrae, the intercentrum is reduced and thepleurocentrum is well developed with broad, expanded and well developed neural arches.The sacrum has two vertebrae.
3. Presence of long interclavicle and expanded ilium.
4. Humerus with a foramen
5. The limbs are pentadactyl in nature with five fingers and toes.
6. The phalangeal formula is 2-3-4-5-4, especiallyin the hind limb skeleton.

Thus seymouria was the basic member among the cotylosaurs. *Hylonomus* from early Pennsylvanian was comparitively small, most primitive and the earliest reptile. *Limnoscelis* (Figure 9.3) was another primitive but larger and heavily built reptile from lower Permian with many typical cotylosaurian characters.

Figure 9.3 *Limnoscelis*

Captorhinus was a small Permian genus about a foot in length. All these primitive cotylosaurs are included in the group Captorhinomorpha. They were probably carnivorous animals that preyed on small amphibians and reptiles as well as large insects. From the ancestral stock like *Hylonomus* or Captorhinus, the cotylosaurs evolved in different directions during late Palaeozoic and early Mesozoic times. One of the early offshoots from the cotylosaurian stock was *Procolophon* of late Permian and Triassic. These were small reptiles similar to many modern lizards in size and habits. They showed a trend towards notable increase in size during Triassic. Pareiasaurs were closely related to Procolophons from upper Permian and were large in size and heavily built (8–10 feet in length) with short tail. The limbs were very stout and the feet were short and broad so as to support the weight of the animal. They were found to have protective body armour. Another group namely diadectomorphs (*Diadectes*) from the lower Permian beds showed an early trend towards large size. Its skull had a prominent notch like that of the primitive otic notch of labyrinthodonts. They had a mixture of cotylosaur amphibian and primitive seymourian traits.

Mesosaurs were the earliest known aquatic reptiles which appeared at the time when the first cotylosaurs established during Pennsylvanian period. *Mesosaurus* (Figure 9.4) was a highly specialized, elongated and small reptile with extended jaws having very long sharp teeth.

Figure 9.4 *Mesosaurus*

It had lateral temporal fossae in the skull but these fossae were different from that of the synapsid temporal openings. It had a long tail and seemed to be adapted for fish-eating. The mesosaurs represent a very ancient and independent group with cotylosaurian ancestry.

An early side branch of the stem reptiles were the chelonians which developed a protective shell and also the ability to pull the head back into the skull. They were able to survive for a long time and are represented today with little or no change.

SYNAPSID REPTILES

Another line in the reptilian evolution was the appearance of remarkable and interesting reptiles namely synapsids. The early synapsids arose during late Pennsylvanian with single lateral temporal opening behind the eye. They bridged the gap between the primitive reptiles and the mammals because they were closely related to the stem reptiles and the latter members approached towards mammalian condition. Pelycosaurs were the first synapsids that appeared in late Pennsylvanian to Permian times. The skull of some early Pelycosaurs resembled the skull of Captorhinomorpha so that it is assumed that they would have had a captorhinomorph ancestry. The first pelycosaurs were the ophiacodonts (Ophiacodontia) of which the genus *Varanosaurus* was a primitive member with specializations in the skull towards the pelycosaurian trend. In general, it was a lizard-like reptile of five feet length with a slender body and limbs and a long tail. It had a narrow, deep and elongated skull, very long jaws provided with numerous sharp teeth and the ears situated very near to the jaw articulaions. The genus *Ophiacodon* (Figure 9.5) was a large Permian reptile which resembled Varanosaurus.

Figure 9.5 *Ophiacodon*

From the ophiacodont stem, the pelycosaurs evolved into two groups namely Sphenacodonts (large aggressive and carnivorous) and Edaphosaurs (large plant-eating). The Sphenacodonts possessed skeleton very similar to the ophiacodonts but they were more advanced in two aspects. In the first aspect, they were adapted for aggressive feeding on other larger animals. As such, they had deep and narrow skull which accomodated the long and strong jaw muscles so that the mouth was able to open widely. The dentition was differentiated in such a way that the premaxilla, front part of the maxilla and the anterior part of the dentary had larger teeth whereas the teeth along the lateral borders were smaller. Secondly, the vertebral spines were highly elongated as in Sphenacodon. In *Dimetrodon*, (Figure 9.6) the spines were greatly elongated to a considerable height at the back in the form of a sail.

Figure 9.6 *Dimetrodon*

However, the significance of these sails were not known. In Edaphosaurs, (Figure 9.7) the skull was short, shallow and remarkably smaller when compared to the size of the body. The teeth were uniformely arranged in a series around the margins of the jaws.

Figure 9.7 *Edaphosaurus*

The vertebral spines were heavier, elongated and ornamented with numerous short irregularly arranged lateral spikes. However, *Casea* and *Cotylorhynchus* were larger and massively built edaphosaurs without sails.

The therapsids or mammal-like reptiles appeared during middle and late Permian and Triassic times from a pelicosaurian ancestry.

In these reptiles, the lateral temporal opening in the skull became enlarged. Quadrate and quadratojugal bones were reduced and loosely connected to the skull which had two occipital condyles as in mammals. They had a secondary palate formed by premaxilla, maxilla and palatine bones separating the nasal passage from the mouth. The dentary bone became enlarged in the lower jaw. In many therapsids, the teeth were differentiated into incisors, canines and molars. The pectoral girdle had two coracoid bones with a large scapula. The ilium was expanded and the sacrum was elongated. The neck was well demarcated from the body. The body was raised from the ground with th help of well developed feet which were adapted for efficient walking and running.

The first formed therapsids were the Phthinosuchians from Permian and Triassic times. The genus *Phthinosuchus* of Permian period was an intermediate form between the late Pelycosaurs and early therapsids. It had skull very similar to that of sphenacodonts but was advanced in having large temporal opening, depression in the jaw articulation and single large canines in both upper jaws and lower jaws. The skeleton was found to be advanced towards theropods. From these intermediate forms, the therapsids evolved into two major groups namely anomodonts and theriodonts.

The anomodont include dinocephalians comprising groups like Titanosuchians, Tapinocephalians, Veryukoviamorphs and Dicynodonts, all of which evolved along several lines. *Titanophoneus* was a medium-sized titanosuchian possessing therapsid as well as primitive characters.

The carnivorous *Jonkeria* and the herbivorous *Moscops* were typical tapinocephalians which were gigantic with a body length of twelve feet or more. The genus *Veryukovia* was intermediate between dinocephalians and dicynodonts with a skull pattern towards dicynodonts. Some dicynodonts like *Dicynodon* and *Endothrodon* from middle Permian were small-sized reptiles whereas the Triassic *Stahleckeria* and *placerias* were massive animals. A lower Traissic Dicynodont namely *Lystrosaurus* was an amphibious reptile.

The theriodonts were the therapsids which evolved in the direction towards the mammals in the middle Permian. They were mostly small to medium-sized reptiles adapted for rapid locomotion. *Cynognathus* of lower Trassic (Figure 9.8) was a typical theriodont with a long and narrow skull and enlarged temporal openings. The teeth were highly specialized and differentiated.

Figure 9.8 *Cynognathus*

The presence of prominent canines was the indicative of its predaceous carnivorous nature. The vertebrae were demarcated into cervical, dorsal, lumbar and sacral vertebrae. The pectoral and pelvic girdles were approaching the mammalian condition. In the limbs, the knee pointed forward and the elbow backward for increased locomotor efficiency. In short, cynognathus seemed to be very close to the mammalian stage. The evolution of therapsids towards mammals was also exemplified by many other theriodonts.

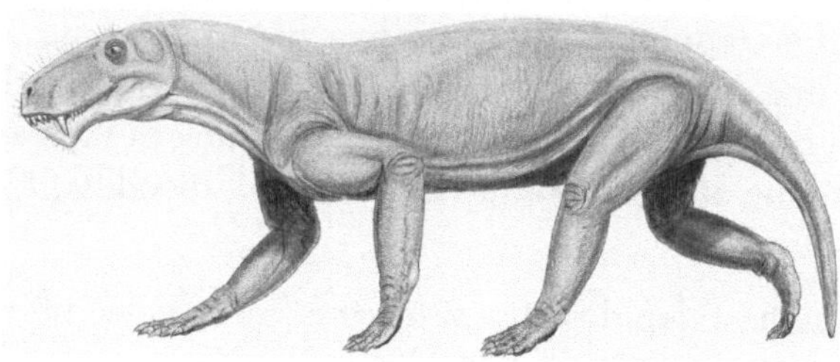

Figure 9.9 *Lycaenops*

The genus *Lycaenops* from Permian time (Figure 9.9) (Gorgonopsians) couldbethedirectancestorsnotonlytoCynognathus but also to another group namely therocephalians (*Lycosuchus*). The tritylodonts like *Tritylodon* were derived from cynognathus during upper Triassic. The other genera similar to Tritylodon were *Bienotherium* and *Oligokyphus* which were considered to be the very early mammals. In all these mammals, the skull had the zygomatic arch which gave attachment to the jaw muscles. In addition, their feeding habits resembled the mammals. The quadrate and articular bones were very much reduced. They had well developed secondary palate. The paired incisors were enlarged and separated from the molars by diastema. The molars had rowed cusps.

Ictidosaurs which comprised the genus *Diarthrognathus* in late Triassic and early to middle Jurassic times bridged the gap between the advanced theriodonts and primitive mammals. In Ictidosaurs, the therapsidan characters reached a high state of perfection towards mammalian condition. They had double jaw articulations namely the ancient reptilian joint between the reduced quadrate and articular and also the mammalian joint between the squamosal and dentary. In this respect, they were true intermediates between reptiles and mammals. It can be mentioned that the ictidosaurs established the base for the origin and diversification of mammals about 100 million years ago.

DIAPSID REPTILES

The diapsids which possess two temporal fossae in the skull, started their evolution during Pennsylvanian period. The first diapsids were the eosuchians (Lepidosauria) which were the founders of the advanced diapsids. The eosuchians gave rise to the other lepidosaurian reptiles including Rhynchocephalia and squamata. Another line of evolution proceeded towards the thecodonts (Archosauria) which appeared at the beginning of the Triassic and gave rise to the other archosaurians such as Crocodilia, Pterosauria and dinosaurs.

The earliest lepidosaurs were the eosuchians of which the genus *Youngina* in the upper Permian period was a lizard-like animal which retained many of the coltylosaurian features. The characteristics of its skull clearly indicate that it would have given rise to the later diapsid reptiles including the modern lizards and snakes. Another primitive lepidosaur namely *Prolacerta* from later Triassic was similar to Youngina.

The thecodonts appeared just prior to the beginning of the Triassic period and all became extinct at the end of Triassic. They comprised four subgroups namely Proterosuchia, Pseudosuchia, Aetosauria and Phytosauria. The protosuchians represented by the genus *Erythrosuchus* from lower Triassic were the earliest and most primitive thecodonts. *Erythrosuchus* was a quadrupedal reptile with the skull retaining a number of primitive features. The only archosaurian feature found in this animal was the pubis which was slightly elongated and highly downturned. But the archosaurian features began to appear inpseudosuchians. *Euparkeria* from the lower Triassic time was the basic pseudosuchian which was a small carnivorous reptile (2 feet in length) with delicate and hollow bones. It was partially bipedal in nature. In slow movements, it used all the four limbs whereas the hindlimbs were used for running by lifting the upper part of the body above the ground by keeping the forelimbs in a free swinging position. Its long tail was used to counterbalance the weight of the body. It had a light, narrow and deep skull with a very large anteroorbital openings in front of the large eyes and two temporal fossae. The presence of large openings at the sides of the lower jaws was one of the archosaurian characters in this animal. The margins of the jaws (premaxilla and maxilla of the skull and dentary of the lower jaw) had sharp teeth which were set in sockets (thecodont type). The coracoid and scapula were well developed in the pectoral girdle and the other bones were greatly reduced. The pelvis was the primitive proteosuchian type. But the ilium was expanded and the pubis was directed downward, thus indicating the evolution towards archosaurian type. The upper Triassic genera namely *Ornithosuchus* and *Hesperosuchus* were similar to Euparkeria but they had two rows of armour plates on middle of the back. This feature seemed to be common in pseudosuchians.

The aetosaurs from late Triassic were the later thecodonts in which the development of the armour reached the maximum as seen in the genera *Aetosaurus*, *Stagonolepis*, *Typothorax* and *Desmatosuchus* (Figure 9.10). These aetosaurs had heavy bony armour plates for protection but the armour plates caused an increased body weight.

Figure 9.10 *Desmatosuchus*

They were secondarily quadrupedal. The forelimbs were smaller than the hind limbs, thus showing their origin from the bipedal ancestors namely pseudosuchians. They had small head with weak teeth in their jaws. During Triassic, the phytosaurs evolved leading to a trend towards large size and bacame secondarily quadrupedal like some armoured aetosaurs. They had crocodile-like elongated body as the front part of the skull and lower jaws were elongated with sharp teeth. The nostrils were far back in front of the eyes.

The body was covered byheavy bony scutes. They had very strong limbs of which the forelimbs were smaller than the hind limbs, thus indicating the bipedal ancestry. These were one among the dominant reptiles during Triassic and were highly predaceous and aggressive, thus strikingly resembling the crocodiles in appearance and habits.

The archosaurians which became adapted for flight at the beginning of the Jurassic times were the pterosaurians. Some of them continued through Cretaceous at the end of which they became extinct.

The Jurassic genus *Rhamphorhynchus* (Figure 9.11) was a reptile with two feet in length and the capability of sustained flight. It had an archosaurian skull with two temporal openings. The front part of the skull and jaws were elongated and provided with long and pointed teeth.

Figure 9.11 *Rhamphorhyncus*

The neck was long and flexible and the back was short and solid with ribs between pectoral and pelvic girdles. The coracoid and scapula were strong with an elongated sternum which was indicative of the origin of large pectoral muscles for flight. The forelimb had strong humerus and elongated radius and ulna. The fourth finger was very much elongated to support the wing membrane. The fingers in front of the fourth digit were reduced to small hooks and the fifth finger was lost. The pteroid bone which was a small bone from the wrist supported the wing membranes. All pterosaurs had comparitively small and weak hind limbs and were aerial carnivores with the ability for sustained flight. Various rhamphorhynchoids similar to Rhamphorhynchus evolved during Jurassic times. In these animals, the tail was reduced and almost suppressed. In advanced members of this group, the teeth were also reduced or completely absent and the jaws became bird-like beak. Another group namely Pterodactyloids with the genus *Pteranodon* reached the climax of the pterosuchian evolution. Pteranodon was a giant flying reptile with a wing spread of more than 25 feet and was the last of the pterosaurs. Its body was small and the jaws were modified into long toothless beak. The back of the skull was extended posteriorly into a crest. The flight of pterosaurs could be of soaring type without requiring much energy. In some pterosaurs, hair-like structures were found and was suggestive of an insulating body cover to maintain a constant body temperature. However, they became extinct during the late Cretaceous period as they were unable to compete with the then evolved birds.

The first dinosaurs appeared from thecodonts during late Triassic period and continued as dominant reptiles throughout the Mesozoic era. There were two major groups of dinosaurs namely Saurischia and Ornithischia. Most of the carnivorous saurischians retained the general adaptations of the thecodont ancestors and all were bipedal in nature. But the herbivorous saurischians were secondarily quadrupedal. The saurischians were represented by two groups namely Theropoda and Sauropoda. On the other hand, all ornithischians were herbivores in which the front part of the mouth was in the form of a beak without teeth. But the sides of the jaws possessed teeth which were highly modified for cutting or chewing the plant materials. Though most of the ornithiscians were bipedal, many became quadrupedal and comprised Ornithopoda, Stegosauria, Ankylosauria and Ceratopsia. Thus the dinosaurs evolved in a great variety of diversified groups with varied adaptations. There were many small dinosaurs as well as great giants in both length and weight. The gigantism in dinosaurs could be interpreted in correlation with Mesozoic environments. In Mesozoic times, there was plenty of food supply in the form of vegetation which might have favoured the giant herbivores. In turn, the carnivores which were dependent on the herbivores would have evolved into giants. Thus the Mesozoic era was an 'age of dinosaurs' but changes occured in the environmental conditions including vegetation. Despite the fact that the dinosaurs would have adjusted to such changes for a number of years, they all became extinct due to the failure of adaptive value in a long run.

PARAPSID OR EURYAPSID REPTILES

The parapsid reptiles made their appearance as *protorosaurs* during the Permian period from the early eosuchians. The genus *Araeoscelis* was a small lizard-like reptile with the skull having the upper temporal fossae. The protosaurs underwent several lines of adaptive radiation in Triassic times. *Trilophosaurus* of late Triassic had a very deep skull with jaws provided with broad and chisel-like teeth. The genus *Tanystropheus* with a very much elongated neck would be an ancient lizard rather than a Protosaur. These late Protosaurs became extinct at the end of the Triassic period during which the diapsids especially the archosaurs arose.

The ichthyosaurs were the most highly specialized marine reptiles. They might have evolved from cotylosaurian ancestors or they might have had an diapsidan origin. They exhibited a single pattern of adaptation in their history but evidences are not available to show the relationship of this group with other reptilian lines. The genus *Ichthyosaurus* of Jurassic times (Figure 9.12) was a fish-like reptile with a length of ten feet or more. It had a streamlined body which merged with the back of the elongated head so that the neck was absent.

(a)

Figure 9.12 a) *Ichthyosaurus*

The fleshy dorsal fins and caudal fins were like those of fishes and the legs were modified into paddles. In these reptiles, the vertebrae had lost the interlocking mechanism. The skull was elongated due to the lengthening of jaws. The teeth resembled the primitive cotylosaurians as well as labyrinthodont amphibians. The eyes were large and the nostrils were situated far back on the top of the skull. The limb bones were flattened and the limbs exhibited hyperphalangy and hyperdactyly. The caudal fin was of reversed heterocercal type. Thus the ichthosaurs can be considered as big fishes having adaptations for fast swimming. In the middle of Triassic, the ichthyosaurs evolved into two major groups namely latipinnate ichthyosaurs with wide paddles as in the genus *Mixosaurus* and longipinnate ichthyosaurs with long paddles as found in *Cymbospondylus*.

The protosaurians were the ancestors of two other groups of Mesozoic marine reptiles namely placodonts and sauropterygians which evolved during Triassic to the end of Cretaceous times. The placodonts were adapted to live in shallow marine waters at the early Triassic. They had massive stout body, short neck and tail and paddle-like limbs. The nostrils were not terminal but were found immediately in front of the eyes. The strong jaw muscles and the teeth were adapted for feeding on molluscs. The genus *Placodus* was a generalized placodont whereas *Placochelys* and *Henodus* were specialized members with broad, flattened and heavily armoured body. The sauropterygians evolved into two major lines namely Nothosauria and Plesiosauria.

The nothosaurs represented by the genus *Nothosaurus* were small to medium-sized contemporarians of placodonts with elongated neck. The skull was small and the jaws were provided with sharp teeth. The limbs were elongated and strong and the feet were modified into short paddles as in modern seals. The plesiosaurs resembled the nothosaurs but they were specialized in having increased body size and paddles and improved 'fish-catching' jaws. Many Jurassic plesiosaurs attained a length of about 10–20 feet and some Cretaceous members had a length of more than 40 feet. Most of the plesiosaurs had a short and broad body due to the lateral extension of the ribs and a very long neck. The skull was nothosaurian type with large eyes and the nostrils were in front of the eyes. The jaws were provided with very long and sharp teeth for fish catching. The plesiosaurs evolved into two lines namely short-necked plesiosaurs which had short neck and elongated skull as in the genera like *Pliosaurus* (Jurassic), *Trinacromerium* and *Kronosaurus* (Cretaceous) and long-necked plesiosaurs which were larger in size with greatly elongated neck as in the case of *Muraenosaurus* (Jurassic) and *Elasmosaurus* (Cretaceous).

SURVIVING REPTILES

A number of reptilian groups belonging to nearly ten orders were present at the end of Cretaceous period but were represented only by five orders at the beginning of the Tertiary period. Of them, the eosuchians became extinct and gave rise to four modern reptilians namely the Chelonia (turtles), Crocodilia (crocodiles) and Rhynchocephalia (sphenodon) during Triassic times and Squamata (lizards) during Jurassic and snakes during Cretaceous. All these have survived from the early Cenozoic to the present times

The turtles have evolved directly from the cotylosaurs and they retain the solid skull roof. The genus *Eunotosaurus* from Permian period possessed the vertebrae and ribs which were on the development towards the turtle shell and it would be the probable ancestral stock to the Chelonians. The first turtles appeared during the middle or late Triassic times and exhibited adaptations similar to that of the modern turtles. The genus *Proganochelys* of Triassic had basic turtle adaptations with the body covered by a heavy shell and the jaws devoid of teeth. In the later more advanced turtles, the teeth were absent and they developed the ability to pull back the head, limbs and tail into the shell. In terrestrial forms, the limbs were heavy and the feet were short with reduced number of toe bones whereas the short feet were modified into paddles in the marine individuals. The proganochelyids gave rise to amphichelyids which evolved into two groups namely the pleurodires and cryptodires during the late Mesozoic era. In pleurodires, the neck is bent laterally when the head is withdrawn into the shell as in *Podocnemis*. The cryptodires which are highly successful turtles found all over the world, withdraw the head and shell by a S-shaped bending of the neck. These have undergone varied adaptive radiations since Cretaceous times. Some of them are very small (*Testudo*, the modern galapagos tortoise) and some like *Colossochelys* are larger in size. Aquatic turtles live in rivers, marshes and seas and terrestrial forms are found in dry lands, forests, plains and deserts. These chelonians are capable of survival for a very long period due to their structural adaptations to the varied environmental conditions of the world.

The crocodilians are the largest and very aggressive modern reptiles which are closely related to the dinosaurs. *Protosuchus*, *Sphenosuchus* and *Proterochampsa* from the late Triassic times were the early crocodilians. The *proterosuchus* was a small reptile which evolved from thecodonts and well adapted for running on land. The protosuchians gave rise to the mesosuchians which were the most numerous and dominant crocodiles during Mesozoic era from the early Jurassic to the end of Cretaceous. The mesosuchians differed from the protosuchians as well as modern crocodiles in having very large upper temporal fossae for the accomodation of the strong temporal muscles. Many of these early crocodiles were adapted to live along the sea shores and in open oceans as in the case of *Teleosaurus* and *Steneosaurus*. Some like geosaurs were well adapted for marine life with the limbs modified into paddles. The mesosuchians evolved into two advanced crocodilian groups namely *Sebecosuchia* which includes *Sebecus* and *Baurusuchus* having world-wide distribution and the advanced eusuchians namely the crocodiles of modern times. The eusuchians appeared in the Cretaceous period, evolved quickly and displaced the mesosuchians. They were widely distributed in the Cretaceous and Cenozoic times. The genus Phobosuchus of late Cretaceous with a length of 40–50 feet was the largest of all fossil crocodiles. The eusuchians evolved into three major groups of reptiles namely crocodiles, gavials and alligators. The size of these crocodiles is variable ranging from medium to large size. They possess strong and elongated jaws with sharp teeth, short and strong legs with webbed feet, a long tail with powerful muscle and a heavy armour in the form of bony scutes covered by horny plates on the back and at sides. They are well adapted for swimming and aggressive carnivorous mode of feeding. Among them, the crocodiles are narrow-snouted, the alligators wide-snouted and the gavials are very narrow-snouted.

Rhynchocephalians

Rhynchocephalia is one of the four orders of modern reptiles and is now represented by the single genus *Sphenodon*. The rhynchocephalians appeared from the thecodont ancestors at the beginning of Triassic period and spread throughout the world. One group namely Rhynchosaurs were highly adaptive and persisted for a brief period during middle and late Triassic times. The rhynchocephalians were a very much restricted group and continued with a limited evolutionary line from Triassic to till date. The Jurassic genus *Homoeosaur* closely resembles the modern Sphenodon, indicating that the rhynchocephalians have undergone very little change over a period of more than one hundred million years. Therefore, Sphenodon is considered as the 'living fossil' and has only one species namely *Sphenodon punctatus* (Figure 9.13). It is also known as the Tuatara or Hatteria or New Zealand lizard and is the sole surviving rhychocephalian confined to New Zealand. Though the species is at the verge of extinction, it is now very strictly protected.

Figure 9.13 *Sphenodon*

Characteristic features of sphenodon It is a large lizard-like modern reptile of about two feet in length. It has diapsidan skull with two large temporal fossae. The teeth are fused to the jaws. As in all rhynchocephalians, the front of the skull is formed into a type of overhanging 'beak' provided with teeth. It has many primitive characters as well as lacertilian and crocodilian features which are shown as follows.

Primitive characters

a) The skull is diapsid type with two fossae and two complete arcades in which the infratemporal arcade possesses quadratojugal as in eosuchians and Homoeosaurus.
b) The quadrate is immovable and the palate is formed by large vomer, palatine and pterygoid bones.
c) Presence of abdominal ribs and amphicoelous vertebrae with small wedge-shaped hypocentrum between the well developed pleurocentrum.
d) Absence of copulatory organs in males.

Lacertilian characters

a) Presence of large lizard-like body.
b) The skull with parietal foramen to accommodate the pineal eye.

c) Absence of external tympanic membrane.

d) Jaws with acrodont dentition.

e) Presence of T-shaped interclavicle in the pectoral girdle.

f) The heart is incompletely four chambered.

g) Occurrence of cloaca as a transverse slit.

Crocodilian characters

a) Skull with well developed complete infratemporal arcade.

b) Hard palate formed by vomer, palatine and pterygoid bones.

c) Presence of immovable quadrate, proatlas and uncinate processes on the ribs and erect ilia.

On account of its primitive characters, Sphenodon was considered to be the ancestors for other diapsid reptiles. But it is definitely a primitive and generalized type of reptile so that sphenodon cannot be the ancestral stock or degenerate form. It is a persistent type that thrived with very little changes since Jurassic times so that it is a 'living fossil'.

Squamata is the most numerous and highly diversified group comprising the lizards and snakes. *Paliguana* (early Triassic) and *Saurosternon* (upper Permian) were the first formed lizards evolved from eosuchian ancestors. These primitive forms were collectively called as Paliguanids from which two late Triassic lizards namely *Kuehneosaurus* and *Icarosaurus* were evolved. The skull of these lizards retained primitive characters but the thoracic ribs became greatly elongated and supported a membrane for gliding as seen in the present day *Draco*. The genus *Prolacerta* of the early Triassic period evolved as an early lateral branch from eosuchian ancestors like Youngina. It had a cheek similar to the ancestral lizards but the skeleton retained the basic eosuchian characteristics. As the lower temporal arch of this lizard possessed a gap, it is probable that this genus could be the very near ancestor of lizards. From the Triassic times, the lizards became established during the Jurassic period and evolved along multivaried lines of adaptive radiation through the middle and late Mesozoic and Cenozoic times . The snakes are the highly modified lizards and were evolved as last of all the reptiles from the lacertilian ancestors. In snakes, the limbs are absent and the body and tail are greatly elongated with respective modifications in the internal organs. Many have developed highly poisonous venom. The fossil record of snakes is very little so that the evolution of snakes has to be derived from comparing with the modern forms.

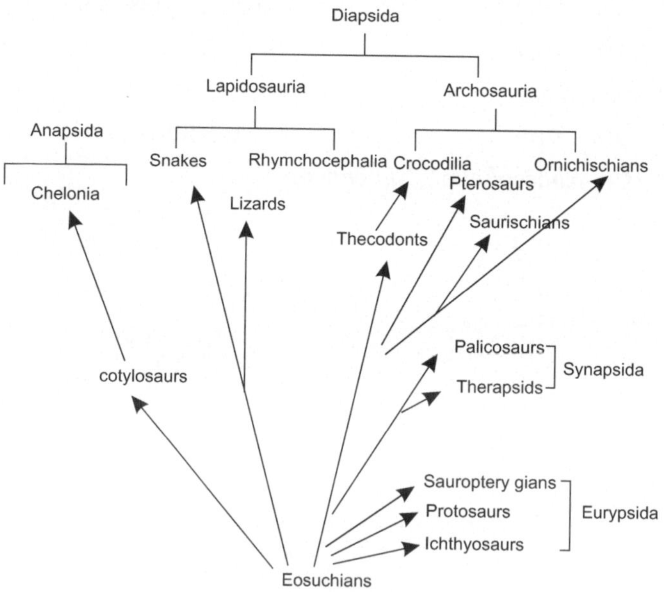

Evolutionary tree of reptiles

QUESTIONS

1. Write a brief account of origin and evolution of reptiles.
2. Explain the fossil anapsids / diapsids / synapsids/ euryapsids with an emphasis on their phylogenetic significance.
3. Write an essay on the importance of Cotylosauria.
4. Describe the important features of Seymouria and add a note on its evolutionary significance.
5. Give an account of pelycosaurian fossils and their role in the evolution of reptiles.
6. Describe the therapsid reptiles with suitable examples.
7. Explain the evolutionary lines of lepidosaurians and archosaurians.

8. Discuss the evolutionary trends of surviving reptiles.
9. Describe the distinguishing features of Sphenodon and its importance in evolution.
10. Write notes on:
    a) *Seymouria*
    b) Cotylosauria
    c) Pelycosaurs
    d) Theriodonts
    e) *Euparkeria*
    f) Pterosaurs
    g) Dinosaurs
    h) Ichthyosaurs
    i) *Rhynchocephalia* / *Sphenodon*

10

DINOSAURS

Dinosaurs were the terrible lizards which lived 200 million years ago for more than one hundred million years during the Mesozoic era. At the time when dinosaurs appeared, the climate was warm and moist and the land was marshy and swampy with plenty of green plants. These conditions favoured them for their great divergence so that they mastered and ruled the earth. The dinosaurs evolved into two major groups namely Saurischia and Ornithischia. In saurischians, the pelvis was a triradiate structure and the teeth were either on the margins of the jaws or limited to the front portion of the jaws. They were carnivorous forms and they used the hind limbs for swift running while the forelimbs were used for procuring food. In ornithischians, the pubic bone of the pelvis was parallel to the ischium as seen in birds and so the name Ornithischia ('bird-like pelvis').

In majority of ornithischians, the ilium was greatly extended with pubis having a large anterior process in the form of a tetraradiate structure (Figure 10.1). These were slow walking animals on their four legs and were carnivorous forms.

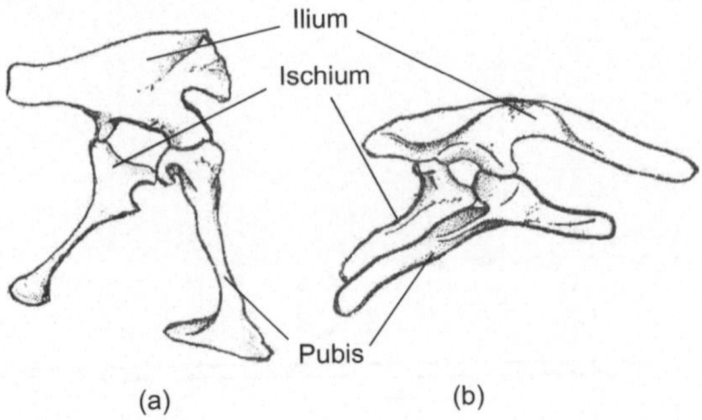

Figure 10.1 Pelvic girdles of dinosaurs

SAURISCHIAN DINOSAURS

The early carnivorous theropod saurischians were well represented by the genus *Coelophysis* from upper Triassic times with a number of primitive thecodont features. It was a bipedal and lightly built reptile measuring eight feet in length with hollow bones. The neck and the tail were long. The skull was long and narrow containing large temporal and preorbital openings. The jaws were sharp, serrated, laterally compressed and set in sockets. It had a saurischian type of pelvis. The genus *Coelurosaurs* resembled Coelophysis and evolved along three distinct lines. In one line, some theropods remained small and retained the primitive ancestral characters. *Ornitholestes* (Figure 10.2) from late Jurassic times was a little coelurosaurian about five or six feet in length.

Figure 10.2 *Ornitholestes*

Though it had very similar characteristics like that of Coelophysis, it was somewhat specialized in having elongate forelimbs, especially the hands with more grasping power. The hand had only three long and flexible fingers with sharp claws. The teeth were adapted to eat all kinds of small animals. The late Cretaceous genus *Ornithomimus* (Figure 10.3) was about the size of an ostrich with a very small skull having no teeth.

Figure 10.3 *Ornithomimus*

The neck was long and jaws were modified into a beak as in ostriches. Therefore, it is otherwise called as 'Ostrich-dinosaur'. Its forelimbs were elongated and enlarged whereas the hind limbs were very long and slender with bird-like feet.

A second line of evolution led to the origin of small to medium-sized theropods like *Deinonychosaurus* during Cretaceous period. Its skull was large and the jaw margins contained serrated teeth. It was bipedal but its fore limbs and hands were very large. In th genus *Deinonychus*, the feet were highly specialized, i.e., while the third and fourth digits were of approximately equal in size, the second digit was much enlarged with large claw for the use as the organ of offence and defence as well as for uplifting the body while walking or running. The tail was provided with stiff long bony rods for effective balancing of the body in locomotion.

The third line was towards the evolution of gigantic dinosaurs such as carnosaurians during the late Jurassic and Cretaceous times. They were the largest carnivores which retained bipedalism as in the ancestral theropods. In these reptiles, bipedalism was strengthened by the presence of very strong and heavy hind limbs with the reduction of forelimbs and hands. In *Allosaurs*, (Figure 10.4) the small clawed forelimbs were used for feeding the food. The *Tyrannosaurus* (Figure 10.5) was a giant dinosaur about 40 feet in length.

Figure 10.4 Allosaurus

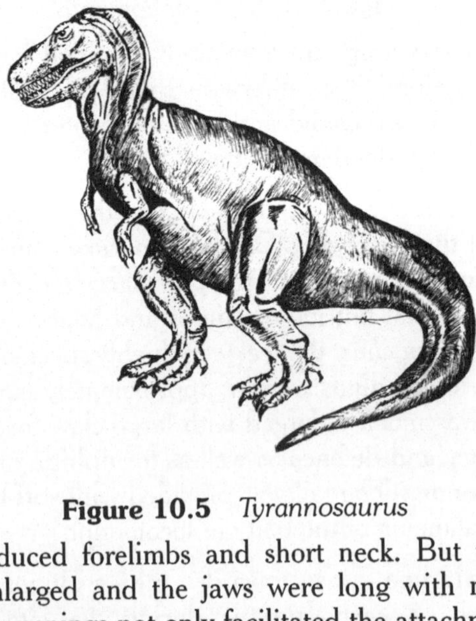

Figure 10.5 *Tyrannosaurus*

It had reduced forelimbs and short neck. But the skull was very much enlarged and the jaws were long with massive teeth. Its large skull openings not only facilitated the attachment of strong jaw muscles but also helped to reduce the weight. Thus they were highly adapted for hunting large animals.

Prosauropods (Palaeopods) were the specialized giant sauropod dinosaurs in Jurassic times. These were probably the primitive saurischians among which the earliest members were smaller in size and the Triassic members were the largest dinosaurs. The genus *Plateosaurus* was 20 feet in length with a small light skull. The teeth were flattened and spatula-like adapted for plant eating. The bones were nothollow and the legs were heavy. The forelimb was secondarily enlarged whereas the hind limb was broad. It was capable of walking on its four legs as well as on the hind legs alone. The prosauropods were mostly herbivores but some were also carnivores.

Some of these small dinosaurs might have given rise to the largest herbivorous sauropods like *Brontosaurus*, (Figure 10.6) *Diplodocus* (Figure 10.7) and *Branchiosaurus* of Jurassic times. These were the largest dinosaurs on the earth with the greatest size having a length of 60–80 feet and the weight of about 30–50 tonnes.

Figure 10.6 *Brontosaurus*

Figure 10.7 *Diplodocus*

They had very long neck and tail. The skull was comparitively smaller and the teeth were present in the front portion of the jaws. The vertebrae were large especially in the neck and back having strong spines for the attachment of muscles. The limbs were very heavy and the bones were dense and solid to support the huge weight of the body. Though these dinosaurs were quadrupedal, their forelimbs were smaller than the hind limbs, thus exemplifying the possible bipedal ancestry. The pelvis was huge and shoulder blades were very long and heavy. They had a very much broad feet with large foot pads and short toes some of which were provided with claws.

ORNITHISCHIAN DINOSAURS

The first known, small-sized and herbivorous ornithischian was *Heterodontosaurus* (Ornithopoda) from the upper. Triassic times. It had a tetraradiate pelvis. The teeth were bsent in the front portion of the mouth whereas the jaws were modified in the form of a beak. The lower front part of the beak was formed by a new single bone called predentary. The teeth were adopted for feeding vegetation and were present only at the sides of the jaws. *Camptosaurus* from upper Jurassic period (Figure 10.8) was a typical and generalized ornithischian from which more specialized ornithopods evolved.

Figure 10.8 *Camptosaurus*

It was a medium-sized bipedal reptile of about six or seven feet in length with a heavy tail. It was a heavier animal than the theropods. The skull was long and the jaws had elevated coronoid process for the attachment of strong temporal muscles. The front part of the jaws had no teeth but broad leaf-like teeth were present at the sides of the jaws. The premaxillary bones were broadened to form a flat beak. When the mouth was closed, the cheek teeth were brought together like a crushing mill.

The fore limbs were strong and the fingers in the hands were short and provided with blunt nails. The hind limbs were also strong with the feet having forwardly directed four toes only. The pelvis was ornithischian type. *Hypsilophodon* was another camptosaurid ornithopod in lower Cretaceous times and was considered to be more primitive than Camptosaurus. It was a very small dinosaur with a lightly built body, long tail and long-toed hind limbs. The presence of teeth in the premaxillary bones is a primitive character in this reptile.

From the camptosaurid-like ancestors, the ornithopod dinosaurs diverged into three lines. In one group, there was continuation of the camprosaurid characters and a trend towards increased size in some genera. For example, *Iguanodon* from the lower Cretaceous was a very big dinosaur with a length of more than 30 feet. Along the second line, small dinosaurs such as *Stegoceros* and large sized *Pachycephalosaurus* were evolved. In these reptiles, the skull roof was very thick so that the upper temporal opening was completely eliminated and the lateral opening was suppressed.

The third line of evolution was towards the most spectacular ornithopods namely Trachodonts or Hadrosaurs of late Cretaceous period. The genus *Anatosaurus* was a large sized aquatic dinosaur with a length of 30–40 feet. The front part of the skull and the lower jaw were broad and flat resembling the bill of a duck. Therefore, they were commonly called as duck-billed dinosaurs. The jaws were provided with 500 or more number of teeth which were used for crushing the food. In these animals, the premaxillary and nasal bones were drawn over the top of the skull to form a hollow crest through which the nasal passage traversed. In the genus *Corythosaurus*, the crest was helmet-shaped whereas it was hatchet-shaped in *Lambeosaurus* and long tubular in *Parasaurolophus*. This modification might have helped to increase the sense of smell and to enhance the power of voice. The duck-billed dinosaurs were bipedal reptiles having very heavy hind limbs with broad webbed feet.

The stegosaurs or plated dinosaurs were very large and earliest known camptosaur ornithischians from the early Jurassic times. They became secondarily quadrupedal with peculiar plates on the back and spines on the tail. The upper Jurassic *Stegosaurus* (Figure 10.9) measured about 20 feet or more in length with a small skull as in camptosaurids. The presence of much smaller brain with an enlarged spinal nerve in the sacral region led to the belief that they were double-brained dinosaurs.

Figure 10.9 *Stegosaurus*

It was a quadrupedal animal but the hind limbs were much larger than the forelimbs. There were two rows of alternately arranged vertical and triangular bony plates on the middle of the back. The tail had four long bony spikes as protective structures. These bony plates were believed to be either protective or the structures of species and sexual recognition. But the recent interpretation is that the plates could have been used to control temperature, thus acting as thermoregulators. The Stegosaurs were the first large dinosaurs to become extinct during late Cretaceous.

The ankylosaurs were the armoured ornithischian dinosaurs of Cretaceous times. The genus *Ankylosaurus* (Figure 10.10) was a bulky quadrupedal dinosaur with 20 feet in length. Its skull was very broad and the teeth were small and weak.

Figure 10.10 *Ankylosaurus*

The legs were heavy and the hind limbs with short feet were much longer than the forelimbs. The top of the head and the entire back were covered by heavy polygonal bony plates that formed a continuous armour. The sides of the body had long bony spikes. The tail was also armoured ending in a knob formed off by a large bone. These armours were definitely the structures of offence and defence.

The ceratopsians were the horned dinosaurs and were the last to evolve during late Cretaceous. These were the most spectacular and interesting of all dinosaurs. The Cretaceous *Psittacosaurus* was considered as the first primitive and bipedal ceratopsian with short and deep skull. The front portion of the skull was hook-like resembling the beak of the parrot. However, the first true ceratopsians were *Leptoceratops* and *Protoceratops* (Figure 10.11) which were small quadrupedal dinosaurs with secondarily enlarged fore limbs. The feet were broad and adapted for walking.

Figure 10.11 *Protoceratops*

The skull was deep and narrow forming a beak. But it was enlarged by the backward growth of the parietal and squamosal bones which formed a large perforated frill at the back of the skull. The frill in these reptiles could be an enlarged structure to accommodate the strong temporal and neck muscles.

In addition, it might have had protective function and used for the recognition of the species to some extent. The early ceratopsians were hornless but the largest known genera like Protoceratops had small horns on noses. From these Protoceratops, the horned dinosaurs evolved with the trend towards increased size (even up to the length of 25 feet) and development of horns on the skull. In some like Monoclonius, the front of the skull had a single large nasal horn. The genus Styracosaurus had a large nasal horn as well as a series of long bristle-like spikes at the edges of the frill. In genera like *Chasmosaurus*, *Pentaceratops* and *Triceratops* (Figure 10.12), there were two large brown horns above each eye, in addition to the nasal horns. The fenestrae in the frills were secondarily closed in Triceratops which was the last of these reptiles.

The horned dinosaurs were large herbivores similar to rhinoceros and they used the horns as defensive organs and also for the recognition of species and sex.

Figure 10.12 *Triceratops*

Eventhough the Mesozoic era was the 'Age of Dinosaurs', both saurischian and ornithischian dinosaurs appeared only at late Triassic times. While the saurischians are well documented in the fossil record, the fossil record of ornithischians is much more restricted. The dinosaurs failed to cope up the transitional times from the Mesozoic to Cenozoic era so that the regime of dinosaurs came to an end at the end of Mesozoic era. The extinction of these dominant Mesozoic reptiles was a sudden and dramatic one. But the reasons for the extinction of these great reptiles are very difficult to explain ('vide in the topic 'Mesozoic reptiles'). Not only the dinosaurs but also the ichthyosaurs, the plesiosaurs and the pterosaurs disappeared from the earth's history by the beginning of the Cenozoic era.

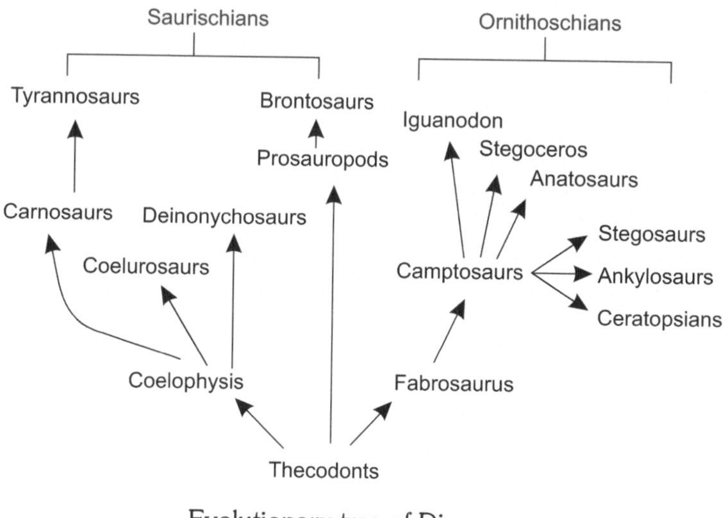

Evolutionary tree of Dinosaurs

QUESTIONS

1. Write an essay on dinosaurs.
2. Differentiate between saurischian and Ornithischian dinosaurs.
3. Classify different groups of dinosaurs with characteristics of the respective representatives.
4. Write notes on :
   a) Saurischia
   b) Ornithischia
   c) *Coelophysis*
   d) Prosauropods
   e) *Brontosaurus*
   f) Camptosaurs
   g) Horned dinosaurs

11

GOLDEN AGE OF REPTILES (MESOZOIC REPTILES)

The Mesozoic era is popularly known as 'the golden age of reptiles' as it was dominated by many remarkable and spectacular reptiles. It was the time during which the great reptilian groups flourised along with the appearance of the existing reptiles as well as the birds and mammals. The cotylosaurian stem reptiles evolved and diversified considerably during Permian period itself. They thrived well during the Mesozoic era and were seen in all parts of the world except New Zealand. They were considered as 'the rulers of the earth' because they invaded all the available habitats (arboreal, aerial, amphibious, aquatic, fossorial and cursorial) due to the absence of tough competitors. The important reptilian groups of the Mesozoic era include the huge dinosaurs and other reptiles which dominated land, the flying reptiles or pterosaurs in air and fish and lizard-like swimming reptiles such as Ichthyosaurs, Plesiosaurs and other marine reptiles in sea. Of all, the dinosaurs were the most spectacular and a great majority of them were giants. Therefore, the Mesozoic era is also known as 'the age of giants' or 'the age of dinosaurs'.

The reptiles made their appearance and invaded the land during Permian period more than 250 million years ago. The Permian land was attractive with well developed vegetation and presence of insects. Moreover, the life in the sea was more competitive so that the reptiles gradually invaded the land with the development of land eggs. During this period, the reptiles diverged into many lines in different times giving rise to the turtles, Ichthyosaurs, mammal-like reptiles and the ruling reptiles namely the archosaurians from which the dinosaurs arose. It is a true fact that the evolution of dinosaurs cannot be compared to any events that occured in earth's history.

MESOZOIC ERA

The Mesozoic era includes Triassic, Jurassic and Cretaceous periods.

The Triassic period was characterized by

- the appearance of the ancestral lizards and turtles and rhynchocephalians,
- the beginning of marine reptiles like Ichthyosaurs and Sauropterygians,
- the abundance and radiation of thecodonts and their extinction, and
- the appearance of dinosaurs with rapid evolution of saurischians and extinction of mammal-like reptiles.

The Jurassic period was characterized by

- the continued expansion of bipedal dinosaurs with the evolution of armoured individuals andsauropods,
- the major radiation of marine reptiles especially Ichthyosaurs and Plesiosaurs, and
- the appearance of crocodiles and pterosaurs.

The Cretaceous period was characterized by

- the appearance of marine lizards and turtles along with gradual reduction of Ichthyosaurs,
- the radiation of crocodiles and turtles with the progress of snake ancestors,
- the climax of pterosaurs,
- the evolution of largest carnivores and the dominance of sauropods,

- the appearance of duck-billed dinosaurs, heavily armoured ankylosaurs and horned dinosaurs, and
- the complete extinction of the most of the Mesozoic reptiles including the dinosaurs,

Ichthyosaurs, Plesiosaurs, marine lizards and pterosaurs (at the end of the Cretaceous period).

Thus the Mesozoic era is characterized in the beginning by the decline of amphibians and the rise of reptiles, by the appearance of birds and mammals in the middle and by the rapid extinction of many reptilian groups at the end.

Decline or Extinction of Mesozoic Reptiles (Causes for the Extinction of Dinosaurs)

The age of reptiles is characterized by three great evolutionary processes namely survival, adaptive radiation and extinction. On earth, there were sixteen orders of reptiles which were adaptedand diversified in such a way that they exploited all the available opportunities of the environment. The extinction of the reptiles, especially that of dinosaurs is one of the most impressive events because the adaptive radiation of dinosaurs was the most spectacular show. By the end of Cretaceous times, most of the Mesozoic reptilian life became vanished leaving a few representatives such as chelonians, crocodiles, Rhynchocephalians, lizards and snakes. It was not a single species of reptiles that disappeared but several reptile orders completely disappeared. Some were tiny reptiles, some were gigantic, some were moderate and some were spectacular animals that ever lived on the earth.

The disappearance of a number of vertebrate groups especially the dinosaurs 65 million years ago at the Cretaceous–Tertiary boundary is referred to as KT event. According to Benton (1993), there was a mass extinction of ranges of vertebrates across the KT interval and the highest extinction rates were met within dinosaurs, Pterosaurs and Plesiosaurs (all 100%) as shown in the following table.

Table 11.1 Extinction of vertebrates at the KT boundary

Vertebrate group	No. of surviving families	No. of families which became extinct	Rate of extinction (%)
Cartilagenous fishes	44	8	18
Bony fishes	50	6	12
Amphibians	11	0	0
Reptiles	71	36	51
Turtles	15	4	27
Lizards and Snakes	16	1	6
Crocodiles	14	5	36
Pterosaurs	2	2	100
Dinosaurs	21	21	100
Plesiosaurs	3	3	100
Birds	12	9	75
Mammals	22	5	23

Some groups of vertebrates disappeared at the KT boundary itself whereas others became extinct before the end of Cretaceous period. But the exact causes for the decline of these reptilian groups are not clearly known. However, a number of theories have been made available over the years with regard to the extinction of the dinosaurs. Any how, the following reasons could be attributed as the possible causes for the extinction of the Mesozoic reptiles.

Geological or cosmic disaster According to one view, a major volcanic eruption or the after effects of a major extra-terrestrial impact on earth might have wiped out the giant reptiles. But the amphibians, small reptiles, birds and mammals were not affected by these disasters.

Prevalence of epidemic It is believed that the widespread outbreak of diseases could be the cause of the extinction of the ruling reptiles. However, the other small reptiles were not affected so that this opinion is not accepted.

Racial senescence A common concept is that the dinosaurs died out simply because of racial senility (old age of the race), i.e., the exhaustion of the genetic potential due to the gigantism, excessive spinocity and loss of ability to adapt. But the proper reason is not known for the success of contemporaries like amphibians and the surviving reptiles of today.

Change in diet During Cretaceous period, the swamps and streams were limited by the elevated land mass as the rocks and mountains were coming up. Cosequently, the dinosaurs such as duck bills and sauropods might have restricted in space. In addition, ferns which were the dominant plants during Cretaceous times were replaced by the flowering plants. Therefore, most of the herbivorous reptiles which were dependent on the ferns were not able to tolerate the floral change. Thus the herbivorous reptiles might have become extinct at the beginning followed by the carnivorous dinosaurs. However, this view does not provide answer for the disappearance of marine and flying reptiles.

Predators It is believed that the blossoming mammals could have devoured the eggs and young ones of the reptiles. So, they might have disappeared due to the inability to tolerate the predatory mammals. But the dinosaurs and mammals were found to co - exist from Triassic to Cretaceous times. Moreover this concept does not explain the reason for the extinction of marine and flying reptiles.

Huge size A number of large sized carnivorous reptiles might have preyed the other dinosaurs which became extinct in later period due to the disappearance of herbivores. The large size of the dinosaurs could have been an obstacle to escape from the enemies. The large-sized dinosaurs seemed to have slow rate of breeding with slow growth rate in the young ones.

Constipation and poisoness alkaloids Constipation could be attributed as another possibility for the disappearance of large reptiles due to the scarcity of vegetation and other food materials. In addition, the angiosperms containing poisoneous alkaloids could have restricted the life of herbivorous dinosaurs.

Climatic changes The current scientific hypothesis points that the decline of many groups of large reptiles at the KT boundary is due to long term climatic changes by which the dinosaurian habitats gave way to mammalian habitats. It is evidenced that there were climatic rhythms throughout the geological history of the earth. It is known that the Mesozoic era was subjected to relatively high temperature whereas the Cenozoic era had comparitively low temperature. The late Cretaceous period was known to have colder climate. These climatic changes could be unfavourable to the reptiles which thrived well in warm temperature. Again, the huge reptiles might have become stiff and inactive in cold conditions. In addition, cooler environmental situation would have been unfavourable for the development of the reptilian eggs.

It is also known that the temperature changes were very little in day and night of all seasons during Mesozoic era. But the climatic condition might have changed due to daily and seasonal fluctuations intemperature at the end of Cretaceous times. Therefore, the cold blooded large reptiles that were acclimatized to less temperature variations would have faced impaired regulation of body temperature at the KT boundary. The reduced temperature in the sea water at the end of Cretaceous times could be the reason for the extinction of the large marine reptiles like Ichthyosaurs, Mosasaurs and Plesiosaurs. Thus the events on land and in the sea during the KT boundary provide ample evidence for the decline of many groups of reptiles through late Cretaceous times. While the large-sized reptiles encountered much difficult temperature regulation, the small reptiles were able to regulate their body temperature in one way or other. Large crocodiles and turtles would have escaped from the climatic changes by inhabiting the non-extreme climates. For example, the crocodiles and large turtles were primarily tropical whereas the small turtles occupied the far North and South. The large terrestrial turtles invaded the equatorial islands.

Thus many believe and accept that the climatic changes would be the main cause for the disappearance of reptiles during the KT event throughout the world. However, the reason for the disappearance of Pterosaurs in air has not been unravelled. Though a number of theories ranging from physiological to ecological, from climatic to terrestrial and from topographic to extra-terrestrial concepts are prevalent, the exact reason for the extinction of the Mesozoic reptiles still remains as a mystery.

QUESTIONS

1. Write an essay on 'golden age of reptiles'/Mesozoic reptiles.
2. Give an account of important characteristics of the Mesozoic era.
3. Enumerate the possible reasons for the extinction of the Mesozoic reptiles/dinosaurs.
4. What you know about KT event/Cretaceous–Tertiary boundary.
5. Write notes on :
   a) Golden age of reptiles
   b) Mesozoic era
   c) Climate and extinction
   d) KT event

12

EVOLUTION OF BIRDS

The birds comprise a large group of highly successful flying vertebrates. The problem of the origin of birds has been one of the great challenges in evolutionary biology for a long time. This is because the fossil record of birds is scarce as they were aerial with very delicate skeleton. Undoubtedly the birds have evolved from reptiles and there are great many similarities between reptiles and birds. Therefore, the two groups are together classified as 'Sauropoda'. The birds exhibit more similarity to the diapsidan archosaurian reptiles and so they are considered as archosaurians with feathers and adaptations for flight. According to Huxley, the feathers of birds are nothing but highly modified scales so that the birds are referred to as 'glorified reptiles'.

ORIGIN AND EVOLUTION

It is believed that the birds would have evolved from archosaurian reptiles. Among the archosaurs, the diapsid pseudosuchians are considered to be the avian ancestors. The lower Jurassic pseudosuchian genus *Euparkeria* shows a trend towards avian features. The first known Jurassic fossil birds were *Archaeopteryx* and *Archaeornis* which were discovered from Bavaria (Germany) in 1861. These fossil birds resemble both a bird and a reptile, thus forming a 'connecting link' or 'missing link' between the thecodont Euparkeria and the modern birds.

Archaeopteryx

Archeopteryx (Figure 12.1 and 12.2) was a medium-sized bird like a crow. It had a lightly built skull with a movable quadrate. The jaws had several small widely-spaced sharp teeth in sockets. The neck was curved as in therapod dinosaurs. The tail was long and straight with caudal vertebrae. The cervical and thoracic vertebrae and the pubis contained air spaces. The pelvis and hind limbs were therapodan type. The hand had three elongated fingers bearing long curved claws. The forelimbs could have been used for climbing tree trunks whereas the hind limbs and pelvis were used for moving about on the ground. The wings were similar to the modern birds with 11–12 primary flight feathers attached to the hand and 12 secondaries attached to the ulna. The tail feathers were 16–17 pairs in number and were attached to the caudal vertebrae.

Figure 12.1 A well preserved fossil *Archaeopteryx*

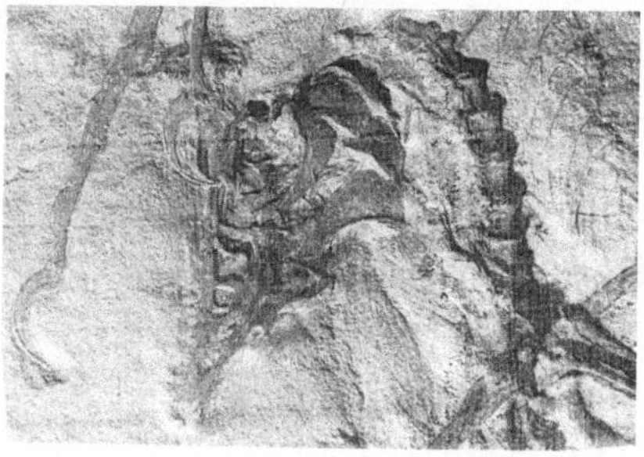

Figure 12.2 A fossil *Archaeopteryx* skull

Reptilian characteristics of Archeopteryx

- Archosaurian skull with two posterior temporal fossae.
- Homodont and thecodont teeth in the jaws as in crocodiles.
- Presence of scales on the feet and base of the beak.
- Presence of heavy non-pneumatic bones.
- Elongated vertebral column with free amphicoelous vertebrae as in sphenodon.
- Presence of 10 cervical vertebrae in the neck.
- Long tapering tail with 20–33 caudal vertebrae.
- Absence of pygostyle.
- Presence of abdominal ribs in addition to cervical and thoracic ribs as in crocodiles and sphenodon.
- Absence of uncinate processes in the ribs.
- Presence of ill-developed keel in the sternum.
- Well developed and free radius and ulna in the forelimb skeleton.
- Presence of free carpals and metacarpals.
- The hand with three fingers provided with claws.
- Therapod dinosaurian hind limb and foot with separate tibia and fibula.
- Presence of simple brain with long and narrow cerebrum and small cerebellum.

Avian characteristics of Archeopteryx

- Crow-like body with long flexible neck.

- Presence of beak, wings, tail and contour feathers.
- The three-fingured forelimbs modified into wings with 6–7 well developed primaries.
- Presence of four digits in the hind limbs with avian phalanges.
- Presence of round skull formed by completely fused bones without sutures.
- Presence of single occipital condyle.
- Eyes provided with sclerotic bones and pecten.
- Pectoral girdle with V-shaped furcula, fused coracoid and scapula and weak sternum.
- The pelvic girdle, bones of upper and forearm and the hind limb skeleton were avian in nature.

Nearly eight skeletons of Archeopteryx are known from the upper Jurassic lime stones Bavaria in the form of skeletons, fragments and imprints of feathers. These birds had retained a number of reptilian characters. Evidences indicate that they were not active fliers but they might have fluttered or glided from tree to tree or ascended for moderate distances to a height.

Views on Evolution of Birds

Thus the bipedal thecodont Pseudosuchians could be the ancestors of birds. As fossil evidences of birds are scanty, many views have been put forward to trace the possible ancestor of birds among the thecodont reptiles.

1. According to one view, the birds might have evolved from pterosaurs as these flying vertebrates resembled the birds in possessing bipedalism, carnivorous habit, thecodont dentition, eyes with sclerotic plates and so on. But this theory has not been accepted on many grounds especially the presence of membranous wings supported by single finger in pterosaurs.
2. Another concept proposes that the dinosaurs possibly the saurischian dinosaurs could be the ancestors from which the birds evolved. This is because these dinosaurs had a number of avian features such as diapsid skull, bipedalism, pneumatic bones, hind limbs with fused bones, forelimbs

with three digits, restriction of teeth, etc. However, this view has not been accepted as the dinosaurs had a number of advanced characters.

3. Another theory considers that the archosaurians were the close relatives of birds and they would have given rise to the reptiles and birds. This is because the archosaurians were bipedal in nature with three digits including one opposable thumb in the limbs. They had delicate, hollow bones and beak-like jaws without teeth. This theory also maintains that the archosaurian ancestral stock could have evolved into many lines, i.e., one line leading to lepidosaurians, another to other archosaurians and still other line to birds as an independent offshoot. Thus these ancient bipedal thecodont archosaurians might have given rise to birds directly, eventhough there is a large gap between reptiles and Archeopteryx. Hence the birds can be considered as 'glorified archosaurs' rather than 'glorified reptiles'. However, the exact ancestral archosaurian reptile is not known and the reason for the transformation of the cold blooded terrestrial archosaurs into warm blooded aerial birds also remains as a mystery.

True birds are believed to be evolved during Cretaceous times or during the early Cenozoic era. During the transitional period of their origin (Triassic period), many changes would have occurred in the endoskeleton along with other specializations for cursorial life. A few fossil birds represent the intermediates between the reptiles and Archeopteryx as well as between Archeopteryx and modern birds. A fossil genus *Sinosauropteryx prima* was a theropod having the body covered with primitive feathers. But it was structurally similar to unfeathered dinosaurs. Ornithominosaurs, Therizinosaurs and Oviraptorosaurs possessed several bird-like features such as pneumatized skull and vertebrae, elongated wings with feathers and so on. Deinonychosaurs which included *Sinovenator*, *Byronosaurus*, *Microraptor*, *Dromaeosaurus* and *Protarcheopteryx* were the most primitive and closest known theropods to Archeopteryx. These had bird-like skull and teeth with the body covered by a variety of feathers including wings, legs and tails. The fossil birds like *Shenzhouraptor*, *Rahonavis*, *Yandangornis* and *Jixiangornis* were slightly more advanced than *Archeopteryx* especially in vertebrae, sternum and wing bones. The first birds are represented by a variety of fossils such as *Sapeornis*, *Omnivoropteryx*, *Confuciusornis* and *Changehengornis*. All these possessed large pygostyle, seven sacral vertebrae, sternum with a keel and reversed hallux. Enantiornithines like *Sinornis*, *Gobipteryx* and *Protopteryx* showed several bird-like features such as fewer dorsal vertebrae, narrow V-shaped furcula, reduction of digital bones in wings, etc. The genera like *Eoalulavis*, *Iberomesornis*, *Cathayornis*, *Patagopteryx* and *Apsaravis* had bird-like characters particularly in the vertebrae and sternal keel. The genus *Mononykus* is a recent find and it is at the basal position in the avian family tree. It had the size of a turkey with the skullhaving tiny teeth, elongated neck and very short forelimbs. This bird could have become secondarily fligtless with several avian characteristics. *Ichthyornis*, *Hesperornis*, *Dinornis*, *Aepyornis*, *Limenavis*, etc. were the advanced flightless birds with more avian characters. Thus the fossil birds include both flying as well as flightless birds.

Diphyletic Origin of Birds

Lowe proposed a diphyletic origin for the appearance and diversification of birds. According to him, the flying birds might have evolved from ancestors like Archeopteryx and the flightless birds would have arisen from ancestral flightless birds independently. Gregory also gave much emphasize to the diphyletic origin of birds, i.e., some of the ancestral birds were acclimatized to cursorial land life and some others became adapted for arboreal life. The modern Carinatae birds would have evolved from the arboreal ancestral stock and the present day flightless birds arose from the cursorial forms.

Monophyletic Origin of Birds

According to Romer, all birds could have evolved from a single ancestor like Archeopteryx and the flightless birds might have arisen from the winged ancestors by the loss of flight secondarily. The time at which the birds originated could be characterized by a heavy competition between the ground-living birds and the newly evolved mammals during the early Cenozoic era. Therefore, a few cursorial birds would have evolved into flying birds by acquiring perfect flight adaptations in order to escape from the enemies. Then they might have diversified with different habitats with modifications in beak and feet. A few others might have remained as flightless birds by the loss of flight in the areas where there were no threat of enemies.

This view has been supported by the ostrich-like flightless fossil bird, *Eleutheornis* which has a close affinity with the modern flying birds and flightless birds having vestigial wings.

ORIGIN OF FLIGHT IN BIRDS

Though it is an established fact that the birds have originated from the reptiles, the exact nature of the ancestral bird and the mode of acquisition of flight mechanism by the birds are not satisfactory. Thus the origin of bird flight is highly speculative and many theories have been put forth regarding the origin of flight in birds. Among them, the following are the important ones.

1. The Arboreal Hypothesis

This theory was given by Marsh (1880) and then elaborated by Hurst (1892) and Pycraft (1896). According to them, the bird ancestor was an arboreal reptile-like animal with four legs capable of parachute mechanism to leap from tree to tree or from trees to the ground. In due course, the gliding flight evolved into powered flapping flight with development of wings and flight muscles. This hypothesis has not been accepted because the ancestors of birds namely the early thecodonts were bipedal cursorial animals.

2. The Cursorial Hypothesis

This theory was proposed by Williston (1879), later elaborated by Nobsca (1923) and supported by Ostron (1974). This hypothesis envisages that the ancestor of birds was a small, long-tailed and bipedal running animal. While the hind limbs were used for running on the ground, the free forelimbs were employed for maintaining balance and for flapping in air to increase the speed. In the course of time, the forelimbs were gradually expanded and modified into wings and the scales became modified into feathers of various types. Thus the primitive birds might have had the ability for short flights initially and then long flights at later stages. In other words, the bipedal locomotion would have transformed into flight in birds.

As per the opinion of Ostrom, the wings of the ancestral birds could have evolved as prey-catching device (insects and other prey) and later became gradually modified into flapping appendages for gliding in air. However, Rayner (1988) raises the point of view to show whether the evolution of flapping in a running animal is for insect catching or for assisted leaping.

3. The climbing Escape Hypothesis

This theory is put forward by Elzanowski (2002) and it narrates that the feathered ancestral bird might have foraged on the ground and perched on a tree or on a rock in order to escape from the enemies. Then it glided down to the ground by spreading its wing feathers to get the lift from the surface. Then the flapping could have evolved to extend the length of flight to the ground.

But all the theories with regard to the origin of flight in birds have their own merits and demerits. However, as Rayner (2001) has stated, transition from a non-flying form to a glider and then to a bird-like Archeopteryx seems to be acceptable. The flight features and wings of Archeopteryx indicate that it was undoubtedly a flier because the forelimbs were modified into wings as in modern flying birds. Moreover, its robust furcula had the pectoralis muscle and the asymmetrical curved feathers allowed aerodynamic mechanism for the wing beat. Elzanowski has also viewed that the flight muscles of Archeopteryx would have adequate power for take-off from the ground and for sustained flapping flight.

ADAPTATIONS ACQUIRED BY MODERN BIRDS FOR FLIGHT

Birds resemble reptiles and at the same time, they are highly modified and superior of them. There are a number of anatomical, embryological, physiological and palaeontological evidences to show the resemblances between birds and reptiles. The birds are highly organized and adapted for aerial mode of life. During the evolution of modern birds, both the external and internal organization have been modified for the aerial habit. The following features have been acquired by the birds in correlation with aerial adadptations.

- The spindle-shaped or boat-like body offering least resistance of air.

- Long mobile neck and placement of heavier organs on the lower side and the lighter ones on the upper side in order to maintain centre of gravity.
- Usage of tail as rudder during flight
- Modification of forelimbs into wings for flight and adaptation of hind limbs to support the body onland.
- Large flight feathers (quill feathers) on the wings and the tail for flight and the contour and filoplumes covering the body as insulating mechanism.
- Reduction of back muscles and good development of flight muscles or pectoral muscles

  (Pectoralis major, pectoralis minor and coracobrachialis for the upward and downward movement of wings during flight).
- Light and spongy bones, many of which contain air (pneumatic) resulting in the reduction ofspecific gravity and weight of the body.
- Rigid dorsal surface due to the fusion of the vertebrae except the cervical vertebrae, preventing dislocation as well as acting as a firm fulcrum.
- The mobile heterocoelous cervical vertebrae, allowing flexibility to the long neck and the slightly movable pygostyle acting as rudder.
- A single occipital condyle allowing the head to turn freely side to side.
- A well developed keel in the sternum to give attachment for the flight muscles and to act as a strong fulcrum for the wing beat.
- The long and stout coracoids and the flat scapula adapted to resist the inward pressure of downward wing beat.
- The fused clavicles to form 'V' shaped furcula acting as a spring to keep the wings apart.
- The modification of forelimb skeleton for flight and the pelvic girdle and hind limbs for supporting the weight of the body.

- Attachment of solid and spongy lungs to the ribs and the air sacs communicating the air cavities of bones.
- Reduction of specific gravity, buoyancy, internal respiration and temperature regulation of the body by the air sacs.
- Closely packed internal viscera (kidneys in bony sockets, heart in a notch of the liver and crowded and overlapped internal organs.)
- Four chambered heart and complete separation of oxygenated and de-oxygenated blood.
- Larger brains for maintaining the balance and muscular coordination.
- Pecten inside the eyes helping in acute vision and the sclerotic bony plates around the eye adjusting the convexity of the eye ball and protecting the eyes against the pressure caused by the wind in flight.

Thus the birds have developed a number of features for flight in almost every part of their body. They are superior than all the other vertebrates as they have evolved body systems adapted for aerial flight.

QUESTIONS

1. Give an account of fossil birds.
2. Write an essay on the origin and evolution of birds.
3. Explain various theories with regard to the origin of birds.
4. Describe the distinguishing features of Archeopteryx and add a note on its evolutionary importance.
5. Enumerate the theories of origin of flight in birds.
6. What are the major adaptations acquired by the modern birds for flight?
7. Write notes on:
   a) *Archeopteryx* b) Fossil birds c) *Archaeornis*

13

RATITAE
(FLIGHTLESS BIRDS)

The group Ratitae includes large flightless birds which are classified as palaeognathae. These are found in Africa, South America, Australia, New Guinea, New Zealand and Madagascar, thus exhibiting discontinuous distribution. The origin of the flightless birds is not clearly known but there are two views to explain their origin. Romer suggested that the flightless birds would have evolved from flying ancestors. According to him, flight in birds is an adaptation to escape from the enemies so that the flightless birds developed gradually into flightless condition in absence of enemies without necessity for flight. This could be exemplified by the flightless birds Kiwi which occur in New Zealand where carnivorous mammals are absent to feed these birds. Another view emphasizes that they might have evolved from many ancestral stocks independently (polyphyletic origin) and not from a single ancestral group. In many areas, they were eliminated today by the carnivores but they survive in various isolated parts of the world where the predators are absent.

Characteristic Features

1. These are large sized flightless birds with reduced wings and flight muscles.
2. The body is covered by the plumage and the feathers have no hooked barbules.
3. The rectrices, oil glands and syrinx are absent.

4. The skull has a single occipital condyle and the palate is dromaeognathous type.
5. The pygostyle is absent and the ribs possess smaller or no uncinate processes.
6. In the pectoral girdle, the coracoid and scapula are short and fused. The clavicles are small or absent without furcula. The sternum has no keel.
7. The pelvic girdle is of reptilian type resembling dinosaurs.
8. The hind limbs are well developed with long, stout and solid bones but without hind toes and air cavities.
9. Males possess long penis and incubate the eggs.

Classification

The flightless birds are generally classified into the following seven groups.

1. *Struthioniformes (African Ostrich)* It has a very large body (8 feet height) with small head and naked neck. The beak is short, broad and depressed. It is the swiftest runner and can run fast in the sand as it has heavily clawed hoof-like foot with 3rd and 4th toes only. It feeds on seeds and fruits as well as anything that comes in the way, including stones. It can live for a long time without drinking water. Its flesh and fat are edible and the feathers are valuable as ornaments. The eggs are eaten by man and the shells are used to prepare utensils.

Therefore, the Ostrich are domesticated for their valuable products. They are found in the plains and deserts of Africa and Arabia, e.g., *Struthio camelus*.

Figure 13.1 Ostrich

2. *Rheiformes (American Ostrich)* Though it resembles the African Ostrich, it is smaller in size (5 feet height) with head and neck partly covered by feathers. Male possesses syrinx by which a hissing sound is produced. It is also a swift runner with the foot having 3 toes. It feeds on grasses, roots and other vegetation. The feathers are used as brooms. The male is polygamous and incubate the eggs. These Ostriches are confined to South America only. e.g., *Rhea americanus*.

Figure 13.2 *Rhea*

3. *Casuariformes* These flightless birds grow to a height of 5 feet and possess rudimentary wings and 3 clawed toes. Among them, Cassowaries are good runners and swimmers. Their head has a helmet-like knob. The long plumage is used for making rugs and mats. They are omnivorous and nocturnal in habit and can be easily domesticated. The males incubate the eggs. Emus are fast runners and swimmers. They have no projection in the head. They are herbivorous and monogamous. The flesh of these birds is edible and the fat is used to prepre oil. These are found only in Australia, e.g., Cassowaries and Emus

4. *Apterygiformes (kiwi)* It is the modern flightless bird with the body having a size of a hen. Its head and eyes are smaller and the wings and the tails are rudimentary. The beak is long and slender with the nostrils found at the tip. The neck and hind limbs are shorter with the foot provided with 4 toes and very very short hallux. The body is covered by hair-like plumes. The front part ofthe head has many elongated hairs. It is restricted only to New Zealand, e.g., *Apteryx* (*Kiwi*).

Figuer 13.3 Kiwi

5. *Dinornithidea* This group includes the largest extinct birds, the skeleton of which are found in New Zealand. These resembled the Ostriches with the body height of about 12–18 feet. But their head and neck were much longer and the body was heavily built. The feet had 3 toes and the wings are absent. e.g., *Moas* (*Dinornis*)

6. *Aepyornithidea (elephant birds)* These became extinct recently and were restricted to Madagascar. The body is heavily built with a height of 7 feet. The legs were long and stout with four-toed foot. The eggs of these birds were the largest known animal eggs, the shells of which are useful for making utensils. The flesh was edible, e.g., *Aepyornis*

7. Tinamiformes The position of these birds is not clear as they possess a well developed keel in the sternum and also wings and muscles as in Carinatae birds. At the same time, the palate resembles ratitae birds. These are restricted to South America extending to Mexico. These are good runners as well as fliers on stimulation, e.g., *Tinamus*

QUESTIONS

1. Give an account of flightless birds / Ratitae
2. Write notes on :
   a) Flightless birds/Ratitae
   b) *Struthio*/Cassowary

14

EVOLUTION OF MAMMALS

The first fully terrestrial vertebrates were the amniotes which progressed towards two distinct lineages namely sauropsids from which the reptiles and birds evolved and the mammalian synapsid ancestors which gave rise to the mammals. The mammalian evolution denotes the climax of organic evolution as the mammals possess highly specialized nervous system and increased intelligence. They have many marked characteristics such as presence of hair, large cranium, feeding the young ones on milk from mammary glands (hence the name mammals), etc. that are not found in other vertebrate groups. The evolution of mammals was a gradual process that occured approximately 70 million years ago.

The early Permian fossils indicate that one synapsid group namely Pelycosaurs were the most common land vertebrates during Permian times. In the middle of Permian, therapsids descended from the Pelycosaurs and were the dominant land vertebrates. These therapsids underwent a series of stages during Triassic by the development of mammalian characters such as bony palate, lower jaw with dentary as the main bone, mammalian jaw, middle ear and so on. The primitive synapsid reptiles especially the Triassic mammal-like theriodonts bridged the gap between the reptiles and mammals as they had both reptilian and mammalian features. The theriodonts would have probably evolved from the pelycosaurs and were the advanced synapsid group forming the central stock from which the mammals evolved at the end of Triassic times.

Theriodont descendents like Gorgonopsians and Therocephalians possessed certain mammalian features indicating an important step towards the mammalian condition.

The advanced theriodonts such as Cynodonts and Ictidosaurs were regarded as the ancestral mammals mostly found in South and North America, the probable birth places of mammals. Cynodonts first appeared in the late Permian and then several families made their appearance during the Triassic period. The first Cynodonts were *Procynosuchus* of upper Permian with a number of mammalian characters in the skull and lower jaw and *Thrinaxodon* of early Triassic with a number of major changes towards mammalian condition. Later cynodonts like *Cyognathus* (lower and middle Triassic), *Probelesodon* (middle and late Triassic), *Probainognathus* and *Massetognathus* (middle and upper Triassic), etc., exhibited advanced features over the first cynodonts. The genera *Tritylodonts* and *Tritheledonts* (upper Triassic to lower Cretaceous) were very close to mammals but they showed no full transition. The Ictidosaurs were the poorly known small animals that were mammal-like in many respects.

In general, a number of trends could be evident during the transitional stage from mammal-like reptiles to mammals. These organizational modifications included enlargement of the temporal fossae, change from single to double occipital condyles, development of secondary bony palate, gradual increase of dentary at the expense of other jaw bones, dentary-squamosal jaw articulation, appearance of middle ear, dental differentiation (change from homodont to heterodont condition and from polyphyodont to diphyodont condition), appearance of mammalian posture and endothermy, evolution of viviparity, etc. However, the exact mammalian ancestor was not conclusively known among the theriodonts. This is because a few theriodonts were advanced towards the mammalian line in some respects and were also primitive in others. Moreover, the ictidosaurs which were highly advanced towards mammals exhibited several different evolutionary lines. Therefore, the early mammals would have evolved from several groups of mammal-like reptiles (polyphyletic origin). The mammalian origin and evolution can be conveniently studied under three stages which occured during Jurassic and Cretaceous periods of Mesozoic era and Cenozoic times.

I STAGE OF MAMMALIAN EVOLUTION

The first mammals which appeared from mammal-like reptiles during Jurassic times were smaller in size and they continued as minor groups in Jurassic and Cretaceous periods. The earliest known Triassic and Jurassic mammalian fossils are categorized into 5 orders namely Decodonta, Triconodonta, Symmetrodonta, Pantotheria (Trituberculata) and Multituberculata. The decodonts were primitive mammals from upper Jurassic and were known as marganucodonts represented by the genera *Morganucodon* (upper Triassic in Europe) and *Megazostrodon* (upper Triassic in South Africa). These decodonts could have given rise to the monotremes.

The monotremes (Prototheria) are the egg-laying mammals such as *Platypus* or duck bills or *Ornithorhynchus* and spiny ant eaters (*Echidna* and *Tachyglossus*) which inhabit Australia and New Guinea. They lack complete fossil record and they might have evolved during Triassic period, most probably from the decodonts as a side chain. These are the very primitive living mammals that are surviving till date without modifications. Therefore they are considered as 'living fossils'. As these mammals possess both mammalian and reptilian characters, they are the intermediate forms between reptiles and mammals. Thus they represent 'the unfinished mammals'.

The triconodonts represented by the genus *Triconodon* from upper Triassic and lower Cretaceous times were considered to be related to the decodonts especially to Marganucodonts. They became extinct at the end of Jurassic without leaving descendants. The symmetrodonts which evolved as side line during upper Triassic were an unsuccessful group and were related to pantotheres. The pantotherians (Trituberculates) of Jurassic times with the genus *Amphitherium* were the early successful mammals. These were considered either to be the direct ancestors of marsupials and placentals during early Cretaceous time or to be very close to the ancestral stock from which the placentals evolved. The *multituberculates* which included the genus *Plagiaulax* were highly specialized group possessing a very heavy skull with strong zygomatic arches. These arches were helpful to give attachment for powerful muscles to move the lower jaw as an adaptation for herbivorous mode of feeding. They had incisors and molars only and there was a gap in the place of canines. The multituberculates were considered to be the ancestors of herbivorous mammals.

II STAGE OF MAMMALIAN EVOLUTION

In the second stage of evolution, the decodonts, tricodonts, symmetrodonts and pantotheres became extinct at the end of Cretaceous period whereas the multituberculates survived into Cenozoic. Before extinction, the pantotheres gave rise to two important mammalian groups namely marsupials and placentals, during early Cretaceous times. The genus *Pantotheria* could be the direct ancestor of the marsupial and placental mammals. The marsupials or the pouched mammals (Metatheria) are the intermediates between monotremes and placentals but are advanced than the monotremes. These three groups of mammals do not represent phylogenetic series because the marsupials and placentals would have evolved independently from different pantotherians. The Metatherians had a world-wide distribution durig Tertiary period but they became restricted to S. America and Australia due to their inability to compete with more efficient placentals. The marsupials differ from the monotremes in having sebaceous and mammary glands, teats, two-headed ribs and viviparous condition. In addition, separate coracoids, interclavicles and cervical ribs are absent in marsupials. They differ from placentals in having pubis with epipubic bone, marsupial pouch for feeding the young ones with milk, double vagina, cloaca, larger eggs and completion of development of the extremely miniature young ones in the marsupial pouch.

Figure 14.1 Didelphis

Didelphis, the American Opossum (Figure 14.1) is a Cretaceous survivor of modern times representing the ancestral stock. Therefore an Opossum-like primitive mammal would have given rise to the marsupials in America and Australia. Hence Didelphis is considered as a 'living fossil'. The predecessors of Didelphis were *Alphodon* and *Eodelphis* from North America and *Peratherium* from North America and Europe. From the didelphid ancestors, carnivorous marsupials of South America namely borhyaenids (*Borhyaena*) might have evolved during Tertiary period. Some borhyaenids like Thylacosmilus became specialized into large tiger-like marsupials in Pliocene but they disappeared due to the placental competitors from North America. The 'Opossum rat' (*Caenolestes*) (Figure 14.2) is the modern representative of coenolestrids. These marsupials are without marsupial pouch and they exhibit another line of marsupial evolution in South America during Eocene. All the American marsupials represent the polyprotodonts except caenolestrids. The caenolestrids resemble the Australian marsupials namely diprotodonts. Thus the evolutionary history of marsupials indicate that they were restricted only to two continents namely South America and Australia. In Australia, they had a wide range of adaptive radiation as the placental competitors were absent and they continued to the present time as dominant group. The Kangaroo is the most highly specialized marsupial found in Australia. Thus Metatheria includes three groups namely Polyprotodontia, Diprotodontia and Caenolestoideda. Polyprotodonts are more primitive insectivorous or carnivorous marsupials whereas diprotodonts are the largest and most highly specialized herbivores. Caenolestoids seem to be somewhat intermediate between polyprotodonts and diprotodonts.

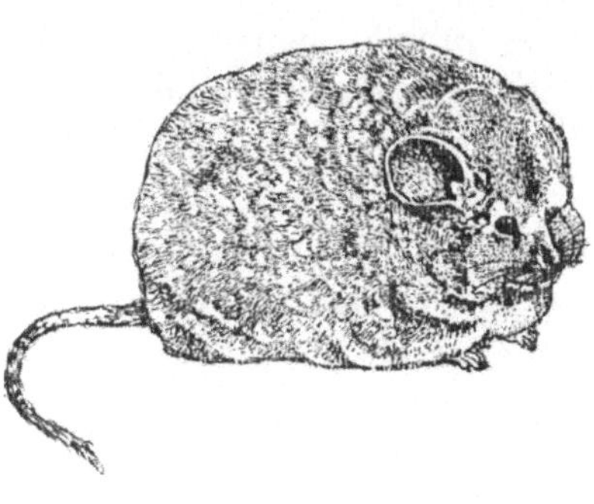

Figure 14.2 Kangaroo rat

During the evolutionary process of mammals, a great parallelism could be observed between the marsupials and placentals in their general adaptations, thus illustrating a good evidence for convergent evolution. The following table illustrates the marsupial and placental genera that exhibit similar adaptations.

Table 14.1 Marsupial and placental genera with similar adaptations

Marsupials	Placentals
Notoryctes	Mole
Dasyurus	Cat
Thylacinus	Dog
The bandicoot	Rabbit
Phalanger	Squirrel
Phascolomys	Bear
Myrmecobius	Chironectes
(Ant eater)	(Otter)
Macropus	Antelope

The group Insectivora has been widely accepted to be the stem placentals from which the eutherian mammals would have evolved through a number of independent lines. A most primitive placental skull from the Cretaceous rocks had many features similar to the shrews (Insectivora). The treeshrew namely *Tupaia* is considered as the most primitive existing placental mammal. All the present day placentals might have evolved as offshoots of ancestral insectivores during Eocene period. The group Rodentia includes the gnawing mammals such as hares, rabbits, flying squirrels, rat - and porcupine - like rodents, etc. These rodents resemble the insectivores more closely (anatomically and embryologically) than any other mammals. These were small-sized, highly successful mammals during Cenozoic times. The evolutionary history of these animals is not fully known because of inadequate fossil evidences. These are more primitive than the present day insectivore survivors so that they might have separated from the placental stock before the appearance of the insectivores. The earliest and most primitive rodents were protogomorphs (the genus *Paramys*). These mammals evolved into numerous lines during Cenozoic period, thus showing a high degree of parallel evolution.

It is not very easy to categorize the mammalian orders since a majority of mammalian groups exhibit isolated positions. Therefore, the mammals could be conveniently divided into different categories based on the general relationships between the groups. According to Gregory, the placentals fall under 4 main assemblages as given in the following table.

Table 14.2 Major categories of placentals and their respective groups

Assemblage	Groups
Insectivore	Insectivora, Dermoptera, Chiroptera and Primates
Edentate	Edentata, Pholidota and Tubulidentata
Carnivore	Carnivora and Cetacea
Ungulate	Perissodactyla, Artiodactyla, Hydracoidea, Proboscidea and Sirenia

Insectivore Assemblage

The group Insectivora includes shrews, hedgehogs, true moles, etc., which are small, primitive and first mammals. As they resemble the stem placentals, they might have evolved first from the placental stock. The primitive Cretaceous insectivores like *Zalambdalestes* were similar to the ancestral insectivores. The dermopteran group represented by a single species *Galeopithecus volans* (*flying lemur*) (Figure 14.3) resembles the shrews of Insectivora, bats of Chiroptera and lemurs of Primates.

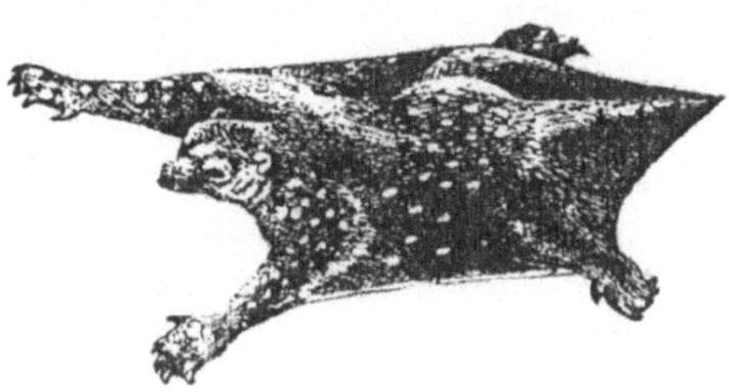

Figure 14.3 *Galaeopithecus*

Chiroptera includes the bats which are the only mammals that have mastered true flight. They would have evolved during late Mesozoic or Tertiary times from the arboreal insectivores which resembled *Galeopithecus*. However, the intermediate stages between the bats and insectivores are not known. The first known bats of Eocene period were typified by *Icaronycteris*. These were highly developed without much variation from the modern forms. The chiropterans include two major groups namely Megachiroptera and Microchiroptera. Megachiroptera comprises large fruit-eating bats and are the most primitive of bats whereas Microchiroptera includes small insectivorous bats. The Vampires are the modern microchiropterans which are adapted for sucking blood from large mammals.

Figure 14.4 Lemur

The primates are the last and most highly evolved mammals comprising a group of more generalized members such as lemurs, tarsiers, monkeys, apes and man. They arose from insectivore ancestors during Cretaceous times, established at the beginning of Cenozoic era and evolved along various lines. Initially, the primitive primates like prosimians (lemurs, lorises and tarsiers) evolved in Palaeocene and Eocene times. (Figure 14.4 and 14.5). The earliest lemurs might have arisen from the arboreal insectivores like *Tupaia*. The primitive fossil primates such as Paromomyids, Carpolestids and Plesiadapids from Europe and North America closely resemble the tarsiers and lemurs. The present day primitive lemurs bear a resemblance to the Eocene *Notharctus*. The anthropoid primates (monkeys, apes and man) are more highly organized than the lemurs and they would have originated from lemuroid ancestors during the upper Eocene and subsequent times. Among them, man is omnivorous, terrestrial, erect, bipedal and cursorial whereas the monkeys and apes are arboreal and frugivorous.

Figure 14.5 Tarsier

The advanced primates (Anthropoidea) can be conveniently divided into three groups namely the New World Monkeys, the Old World Monkeys, the great apes and man and they possess certain common characteristics. The New World monkeys are the most primitive anthropoids found in Central and South America. The fossil history of New World Monkeys is scanty and include the genera like *Homunculus* and *Cebupitheca*. But the fossil record of Old World Monkeys (Cercopithecoids) can be traced from Oligocene to the present. The genus *Mesopithecus* was a Pliocene monkey related to the modern langurs. The Old World monkeys were evolved into two lines namely Cercopithecines (Asiatic and African monkeys like rhesus monkeys, baboons, etc.) and Colobines (langurs). *Parapithecus* was the primitive ancestral cercopithecoid monkey belonging to the Oligocene period. *Propliopithecus* was a very primitive ape which gave rise to *Aegyptopithecus*. These were more advanced than Propliopithecus and were the forerunners of *Dryopithecus* (the ancestors of other apes) and *Limnopithecus* (ancestors of the Gibbons and their relatives). The dryopithecines were the generalized apes very similar to the modern Chimpanzees and gave rise to the apes of Pleistocene and recent times. At the same time, some dryopithecines evolved into a line leading to the ancestors of man. (Refer the topic 'Human Evolution'). The modern apes include Gibbons, Orangutans, Chimpanzees and

Gorillas. *Gigantopithecus* of Pleistocene period were the gigantic apes which were the greatest of all Primates.

Figure 14.6 Chimpanzee

Figure 14.7 Gorilla

Edentate Assemblage

All the three groups included in this category namely Edentata (sloths, armadillos and bears), Pholiodota (ant-eaters) and Tubulidentata (another kind of ant eaters) could have evolved simultaneously along with other groups of mammals from the placental stock during Eocene times. A fossil pangolin namely *Patriomanis* closely resemble the recent ant-eater (Manis) (Figure 14.8). The primitive representatives of Pholidota include Palaeanodonts with the genera *Palaeanodon* and *Metacheiromys* which survived until the end of Oligocene period.

Figure 14.8 Manis

Carnivore Assemblage

The carnivores evolved independently from the ancestral Mesozoic placental stock along with insectivores, rodents, edentates and ungulates. Among the members of Carnivora, the group Creodontia was the most primitive and was regarded as the ancestors of the modern carnivores as well as for the Cetaceans (Whales). The creodonts were the archaic carnivores in Eocene times and they reached the height in Tertiary period. They included two groups namely Okyaenids and Hyaenodontids. These were replaced by the advanced and more specialized carnivores during the transition from Eocene to Oligocene times. All became extinct at the end of Eocene period. One group of creodonts gave rise to the surviving terrestrial Fissipede carnivores such as cats, dogs, bears, wolves, tigers, lions, civet cats, pandas, otters, hyenas, etc. The creodonts also evolved into Pinnipede carnivores like seals, walruses, sea-lions, etc., at the time or soon after the evolution of the early forms. However, they became secondarily adapted to marine life. In addition, large marine mammals such as whales, dolphins and porpoises (Cetacea) were also derived from the Eocene cryodonts.

The cetaceans arose from primitive carnivore-like ungulates known as mesonychids. The whales, after separation from their ancestors, underwent a number of changes and became adapted for marine life during middle Eocene, thus showing "Convergent evolution". The first known fossil whales were known as archaeocetes represented by the genera *Protocetus* and *Eocetus* (middle Eocene) and *Basilosaurus* (Upper Eocene). The modern whales arose during Oligocene from the archaeocetes and almost all families appeared in Miocene times including the small, toothed whales like *Delphinus* (common dolphin) and *Phocaena* (common porpoise). The toothed whales exhibited gigantism as an early trend and the modern sperm whale (*Physeter*) is a giant Odonocete. The whalebone whales (*mysticetes*) are the largest of modern whales. In the later Cenozoic times, the skull of the whales underwent specialization to bear the baleen plates as in *Balaena*, *Balaenoptera* and the titanic blue whales. The blue whales have a length of about 100 feet and the weight of 150 tons, thus exhibiting extreme gigantism in chordate evolution.

Ungulate Assemblage

These are the hoofed mammals (ungulates) which are the last but a great group of placentals. They might have originated from insectivore ancestors during the transitional times between Cretaceous and Tertiary. They seemed to be evolved along 14 lines during Cenozoic era. While some became extinct, others constituted the modern individuals. During Palaeocene and Eocene, a variety of primitive ungulates evolved and began to decline in Eocene. At the same time, the modern forms made their appearance with much diversity and complexity at the beginning of Eocene. The most primitive ungulates were the Condylarths (*Condylartha*) which appeared at the beginning of the Cenozoic time and gave rise to a number of early ungulates (*amblypods*) and South American ungulates (notoungulates, litopterns, astrapotheres, pyrotheres and Xenungulates). In addition, two dominant ungulate groups namely perissodactyles (horses, ases, zebras, rhinoceros, etc.) and artiodactyls (sheep, goat, cattle, swines, hippopotamus, camels, etc.) evolved from Condylarths. The group Perissodactyla comprise the odd-toed hoofed mammals whereas Artiodactyla are the even-toed mammals. Two other groups comprising horses and rhinoceroses also became converged upon a common ancestral stem. The mouse deers were the most primitive of the ruminants and they occupy an intermediate position between the swines and ruminants. Another group namely Proboscidea (elephants, mastodonts, sea cows, etc.,) also evolved independently from the Condylarth ancestors. The elephants are the largest and most highly specialized terrestrial animals. The elephantine characters began to appear in *Moeritherium* which were the earliest proboscidans. The genera like *Palaemastodon*, *Trilophodon*, *Stegodon*, etc., were the transitional forms which approached gradually towards the present condition. The sirenians (Dugongs and Manatees) are distantly related to the elephants and they could be an aquatic offshoot of an early ungulate stock.

III STAGE OF MAMMALIAN EVOLUTION

In the third stage of evolution, the ruling reptiles disappeared at the beginning of Cenozoic era and the marsupials and placentals inhabited various niches, radiated and specialized in various directions.

The mammals exhibited only a low diversity during Mesozoic era for 160 million years. But they underwent major radiation in Palaeocene and early Eocene. These stretches were the best periods for the rapid evolutionary radiation during which more than 20 mammalian lineages arose. Hence, Cenozoic era is considered as the 'Age of Mammals'. But the rate of diversification of mammals has slowed since the Eocene and there is no evolution of major body plans in the last 50 million years.

Thus it is evident that there were a number of independent changes among the members during the evolution and diversification of mammals. In many mammals, the size of the body increased and this enlarged size was advantageous to them in relation to the development of large brain and slow reproduction. In these large mammals, the relative area of heat loss was reduced but they faced the problem of getting the adequate food. They had longer limbs specialized for hand locomotion. In many cases, the number of toes became reduced so that they began to walk on the digits. The number of teeth reduced in them and the teeth also developed additional cusps. The cusps of the teeth were fused to form grinding ridges in herbivorous mammals and cutting blades in Carnivores. Eventhough the earlier mammals possessed brain very similar to those of reptiles, the later individuals had large-sized brain in relation to their behaviour and memory.

QUESTIONS

1. Trace the origin and evolution of mammals.
2. Give an account of mammalian fossils.
3. Monotremes are 'unfinished mammals'. How?
4. In what ways, marsupials occupy an intermediate position between Monotremes and Placentals ?
5. Give a brief note on the trends in the evolution of major mammalian assemblages.
6. Write notes on the following :

   a) Cynodonts b) Pantotheres
   c) *Didelphis* d) Stem placentals/Insectivora
   e) Chiroptera f) Primates
   g) Creodonts h) Ungulates

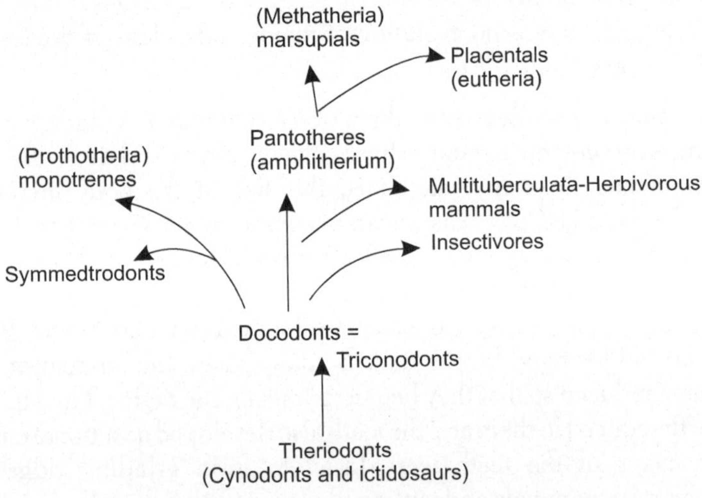

Evolution of monotremes and marsupials

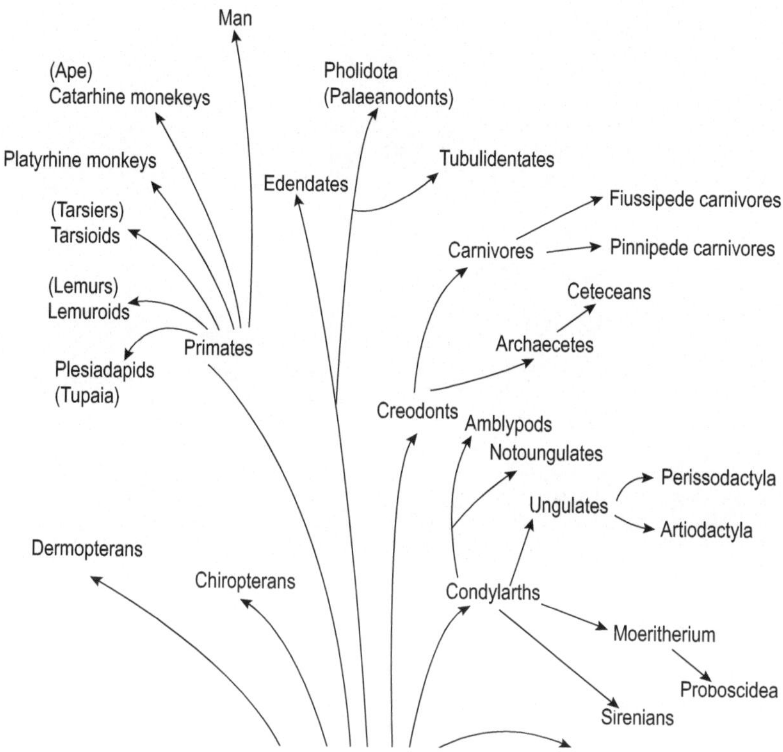

Evolution of placentals

15

MONOTREMES
(Prototheria)

These are primitive egg-laying mammals found in Australia, Tasmania and New Guinea represented by *Ornithorhynchus* and *Echidna*. They might have originated from mammal-like reptiles and their fossils were unknown until Pleistocene period. These are found to persist in all periods of mammalian radiation (Jurassic, Cretaceous and Cenozoic times). *Ornithorhynchus* (Figure 15.1) is otherwise known as duck-billed Platypus and is semiaquatic in habit. It has a duck-like bill and the feet is provided with 5 webbed toes. The body is covered by soft fur. Teeth are absent in the mouth but the mouth has broad bony plates. The tail is long and dorsoventrally flattened and is used as a rudder in water. It feeds on molluscs, crustaceans and worms. *Echidna* is a typical spiny ant-eater (Figure 15.2) with an outer protective covering formed by spines and coarse hairs.

The head is produced into a long and a tapering snout. Teeth are absent and the tongue is very long with a sticky salivary secretion for capture of ants. The tail is very much reduced and the clawed limbs are used to break open the termite nests. It is a nocturnal animal with a tempory marsupial pouch to which a single egg is transferred for a short period of incubation.

Figure 15.1 Ornithorhynchus

Characteristic Features

1. The body has hairs and is provided with sweat and sebaceous glands.
2. The mammary glands are the modified sweat glands and are devoid of teats. The young ones suck the milk from a depression of the surface of the skin glands in which the milk is collected.
3. The ear has three ear ossicles and slightly coiled cochlea but the pinnae are absent.
4. The skull bones are completely fused without sutures and the lower jaw is formed by a single dentary bone.
5. The ribs are single-headed and the cervical vertebrae also bear ribs.
6. The pectoral girdle has a large coracoid with a precoracoid and a T-shaped interclavicle but the scapula is without spine.
7. The pelvic girdle has a pair of epipubic bones.
8. The cerebrum is large and convoluted (*Echidna*) or non-convoluted (*Ornithorhynchus*) without corpus callosum.
9. The aortic arch is single and the RBCs are non-nucleated.
10. The ureters and urinary bladder open seperately into the cloaca to which the penis is attached.
11. The testes and ovaries are abdominal in position. Placenta and vagina are absent in females and they are oviparous.

Figure 15.2 Spiny ant eater

Affinities of Monotremes

The monotremes exhibit a close resemblance to the reptiles from which they have evolved. At the same time, they are undoubtedly mammals as they also possess a number of mammalian characters.

Reptilian Features

1. Tympanic bulla is absent in the skull and true teeth are also absent.
2. The external ear or pinna is absent and the cochlea of the ear is simple and uncoiled.
3. The vertebrae are without epiphyses.
4. The ribs are single headed and the cervical vertebrae possess ribs.
5. Presence of large coracoid with precoracoid and T-shaped interclavicle in the pectoral girdle.
6. Presence of epipubic bones in the pelvic girdle.
7. Absence of corpus callosum in th brain.
8. Presence of cloaca and abdominal position of testes.
9. Oviparous condition and reptilian type of ovaries, oviducts and embryonic development.

Mammalian Features

Monotremes are definitely mammals as they possess mammalian characteristics such as hair, sweat, sebaceous and mammary glands, lower jaw with single dentary bone, three ear ossicles and cochlea in the ear, diaphragm, four optic lobes in the brain, four chambered heart with left aortic arch only, non-nucleated red blood cells, warm-blooded nature, milk secretion and so on.

Thus the monotremes occupy an intermediate position between reptiles and mammals as they resemble both the groups. But they possess more affinity towards mammals than reptiles so that they are considered as 'Unfinished mammals'. These are considered to be 'living fossils' and they could not be the ancestors to higher mammals. They would have evolved as separate line from the ancestral reptiles.

QUESTIONS

1. Write an essay on Monotremes or egg-laying mammals.
2. Give an account of important features of Monotremes and their affinities.
3. Write notes on:
   a) Egg-laying mammals
   b) Monotremes
   c) Echidna
   d) Ornithorhynchus

16

MARSUPIALS

(Metatheria)

Marsupials are the pouched mammals which are more advanced than the monotremes. All are specialized in having an integumentary brood pouch (marsupium) enclosing the teats through which the immature young ones are fed with mother's milk. They are found in varied habits (arboreal, flying and climbing) and majority of them are omnivores. Based on the dentition, the marsupials are classified into two groups namely Polyprotodontia and Diprotodontia.

Polyprotodontia

This group includes the marsupials which have more than 3 incisors in each of the upper jaw and free toes in their legs (didactyly). The American Opossum (Didelphis) is a rat-like, insectivorous and arboreal marsupial without the marsupium. The mother carries the young ones on her back. The native cat of Australia, *Dasyurus* is a carnivorous form with a bushy tail. It is nocturnal in habit with the feet adapted for climbing. *Sarcophilus* is known as Tasmanian Devil which is short-legged and highly predaceous. Chironectes, the water opossum has webbed feet and is piscivorous. *Perameles* are the *Bandicoots* of Australia with pointed snout and long tail about the size of a rabbit and are nocturnal in habit.

Diprotodontia

These are highly specialized marsupials having never more than 3 incisors in the upperjaw. The second and third toes of the hindlimbs are united by a common sheath (syndactyly). *Macropus* (the Kangaroo) (Figure 16.1) is the largest of the living marsupials with short forelimbs and long hind limbs and tail aiding in leaping locomotion. *Phalanger* is an arboreal and noctural with a prehensile tail. *Petaurus* is the flying animal which glides from tree to tree with the help of lateral skin folds between the fore and hind limbs as parachute.

Figure 16.1 Kangaroo

Characteristic Features

1. Presence of marsupial pouch with well developed mammary glands and teats is the most important characteristic feature. In this brood pouch, the immature young ones are fed with the milk of the mother. The pouch is well developed in females but is rudimentary in males. In opossums, it opens backwards but it opens forward in kangaroo. It is absent in Didelphys.
2. The cranial cavity is small and the teeth are 44 in number. The milk dentition is poorly developed.
3. Simple vertebral column with incompletely ossified atlas and with only one sacral vertebrae inmany cases.
4. Presence of double-headed ribs.

5. In the pectoral girdle, the coracoid is much reduced and the interclavicle is absent.
6. The pelvic girdle possesses paired epipulic bones supporting the brood pouch.
7. In the brain, the cerebrum is small with few convolutions but without corpus callosum.
8. The anal and urinogenital openings are separate but are controlled by a common sphincter muscle. The female have two vaginae or three rarely. Fertilization and development are internal but the young ones are born in immature condition.
9. The youngs which are born naked and blind possess clawed fore limbs which help to move them into the brood pouch.

Thus the marsupials possess intermediate features between monotremes and eutherian mammals. They have evolved during Cretaceous time and were widely distributed throughout the world. But they are at present restricted to certain countries (Australia and South America), thus representing a good example for discontinuous distribution. Australia is the major living place of marsupials and they were eliminated in other parts due to the appearance of the placental carnivores in the early Cenozoic era. Today, Australia provides not only a safe home but also supplies abundant food to the marsupials so that they exhibit a wide range of adaptive radiation. In short, it may be said that the marsupials have been an unsuccessful group in the competition with placentals and so they are considered as 'second class mammals'.

QUESTIONS

1. Give an account of marsupials and their importance in the evolution.
2. Classify Metatheria with their characteristic features.
3. Write notes on:
    a) Marsupials/Metatheria/Pouched mammals
    b) Kangaroo

17

HUMAN EVOLUTION

The origin of man is one of the most important events in the evolutionary history of chordates representing the climax of evolution. The placental mammals including the primates might have arisen from the insectivores. The tree shrews are considered to be the first primates evolved from insectivores so that they could be the ancestors for the higher primates which comprise monkeys, apes and man under the division Anthropoidea ('human-like'). The evolutionary history of the primates can be traced back 656 million years ago and the oldest known primate-like mammal was *plesiadapsis* from North America. These animals were found to be widespread in Eurasia and Africa during Palaeocene and Eocene times. The living primates include about 230 species and are characterized by having grasping hands, feet with opposable thumb, flat nails instead of claws, larger brains, closely packed eyes on front of the face with reduced snout, flattened face with stereoscopic vision, parental care of the offsprings and so on.

The advanced primates are grouped as anthropoids which comprise three major categories namely the New World monkeys, the Old World monkeys and the great apes and man. The New World monkeys include the *platyrhines* ('broad-nosed monkeys') which possess broadly-spaced nostrils facing forward and prehensile tail in some genera. The Old World monkeys include the catarhines ('hooked-nose monkeys') which have narrow snouts and non-pehensile tails. Gibbons and Orangutans of Southern and Eastern Asia, Gorillas and Chimpanzees of Africa and humans are included in the group Hominoidea. The living hominoideans include two groups namely Pongidae comprising orangutans and their fossil relatives and Hominidea (the great ape family) with Chimpanzees, gorillas, men and their fossil relatives. The humans are the only extant members of the genus 'Homo' of bipedal Primates.

The origin and evolution of man as a distinct species namely Homo sapiens is referred to as anthropogenesis. The word '*Homo*' means 'Human' derived from the Latin word *humanus*. Anthropogenesis is characterized by a number of important behavioural , morphological, physiological and developmental changes such as bipedalism, larger brain with social learning and language requisition, power of grip, reduced mastigatory system, reduction of canine teeth, hidden estrus, pair-bonding, unique sexual and reproductive behaviours, development of material culture such as physical objects, resources and space, etc. Humans have highly developed brain and are capable of reasoning, language, problem-solving and so on. The mental capability with an erect posture and freely rotatable hands have allowed man to make use of tools.

The human lineage might have diverged by about 5 million years ago from an ape-like ancestor. As the apes exhibit both modern ape characteristics as well as the human features, it is probable that the apes and man would have originated from a common ancestor. Most of the anthropologists assume that humans form a distinct lineage from the great apes on the direct line to men and this split between these two groups could be dated about 15–25 million years ago (late Oligocene or early Miocene). This has been evidenced on the grounds of comparitive anatomy within the fossil record mainly through the intermediate forms or missing links between the modern forms. The closest living relatives of humans are gorillas and chimpanzees but the chimpanzees are considered to be the most closely related species to man. It is estimated that the gorillas would have diverged from chimpanzees about 8 million years ago. Thus it is probable that gorillas and then chimpanzees might have diverged towards the line leading to humans. It is found that the human DNA is identical (98.4%) to that of chimpanzees. By protein sequencing process, the molecular biologists have shown that the humans are much more similar to chimpanzees and they would have branched about 4–9 million years ago. The division of apes and man was confirmed through the existing ape fossils and by the collection of new human fossil specimens. The place of origin of man could probably be Asia or Africa because a good number of fossil apes and primitive men are obtained from Africa, Asia and Europe. A number of fossils have also been found in Central Asia, China, Java and India.

FOSSIL EVIDENCES FOR HUMAN EVOLUTION

The fossil evidences clearly indicate that man and other primates share common ancestors. While the fossil record of the New World monkeys is very scanty, the fossils of the Old World monkeys can be evidenced from Oligocene period to the present times.

The genus *Parapithecus* was one of the first Old World monkeys. They were the anthropoid apes which appeared during Oligocene period. Two lower jaws of these apes were found in Egypt and they formed a link between the tarsoids and Old World monkeys.

Fossil jaws and teeth of ape-like primates belonging to the genus *Propliopithecus* of Oligocene times were found in Egypt (1967) and were considered to be the direct ancestors for modern gibbons.

The Oligocene genus *Aegyptopithecus* from Egypt (1967) was ape-like and resembled *Propliopithecus*. These apes were believed to be the ancestors to *Dryopthecus* which gave rise to gorillas, chimpanzees and man. It is also proposed that *Dryopithecus* was not evolved from *Aegyptopithecus* but they originated from *Propliopithecus*.

Dryopithecus of late Miocene and early Pliocene from East Africa (1930) exhibited a transformation from four-footed condition to bipedalism and erect posture. They had similar size like those of gorillas but their limbs resembled the modern apes and the skulls were similar to that of Old World monkeys. As these fossils looked like chimpanzees, they were regarded to be the immediate ancestors of man. These apes possessed primitive limbs and more specialized teeth than that of man. Therefore, it is assumed that man would have evolved from this group before the specialization of teeth. Thus these apes could be a common stock for men and modern apes and could have given rise to three genera namely *Sivapithecus*, *Gigantopithecus* and *Ramapithecus* about 15–10 million years ago.

The genus *Oreopithecus* of Pliocene times from Italy was similar to man in many features especially in dental nature. They had teeth, jaws, limb bones and face which were very similar to hominoids. So, it is believed that man might have originated from *Oreopithecus* than from *Dryopithecus*.

Ramapithecus from Europe and Asia were man-like fossils and were the contemporians of *Oreopithecus*. Though they were considered to be the perfect intermediate forms between apes and humans, it is shown that they were more closely related to orangutans and were not directly connected to human evolution.

A young anthropoid skull of an ape-man namely *Australopithecus* (Figure 17.1) was found in South Africa in 1925. It had combined characters of ape and man and so it formed a connecting link between the two groups. They had larger brain capacity than gorillas, more rounded forehead and less protruded jaws than chimpanzees, dentition similar to man, indication to erect posture and so on. These characteristics clearly indicate that Australopithecines were the earliest forms with man-like characteristic features leading to human evolution. In 1990s, six species of Australopithecine fossils were recognized and were categorized into two size groups such as lightly-built forms with the brain capacity of 445 cubic centimeter (Australo africanus) and heavier forms with the brain capacity of 520 ccm (*Australo bahrelghazali*, *Paranthopus robustus*, *P. garli*, *P. boisei*, and *P. althiopicus*).

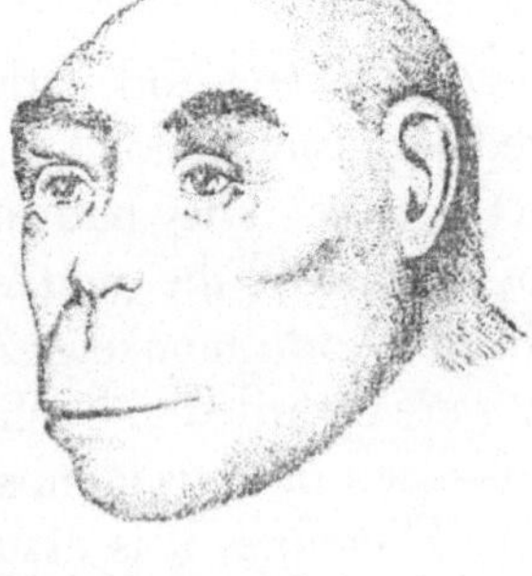

Figure 17.1 *Australopithecus*

HUMAN FOSSILS

The bipedal hominoids emerged during Pleiocene as a unique phase of human evolution. The discovery of man-like fossils revealed that the human evolution had proceeded. The oldest human species was found in East Africa with a large brain capacity (630–700 ccm). They had the hands with the possession of manipulating ability to make tools. They were given the name *Homo habilis* ('handy man'). The earliest man-like fossils namely *Homo habilis* and *Homo rudolfensis* were found in association with the remains of *Australopithecus* in Kenya (1960–63). These differed much from later species of Homo so that they were assigned as *Australopithecus* genus. However, H. *habilis* seems to be having the evolutionary link between Australopithecus americanus and Homo erectus.

A skull of man-like fossil was found in Eastern Java in 1891 and named as *Pithecanthropus erectus* or *Homo erectus* or *Java man* (Figure 17.2). It was considered as the missing link between the apes and man. These fossils had many features resembling the modern man, especially in having the brain capacity of more than 900 ccm.

Figure 17.2 Java man

Therefore, they could be the most primitive human fossils. These were also found in Africa, Europe, Asia and China dating back 1.25–0.5 million years ago with advanced characters over *H. habilis*. They were fully bipedal in nature possessing modern skeleton.

In 1927, a fossil namely *Pithecanthropus pekinensis* or *Sinanthropus pekinensis* or *Peking man* (Figure 17.3) was discovered from Peking.

It is believed that both the Java man and Peking man were closely related.

Figure 17.3 Peking man

A jaw similar to that of man was found in Europe (1907) and given the name *Homo heidelbergensis* (Heidelberg man). These fossils were considered to be the contemporaries of *Pithecanthropus pekinensis*, thus indicating the occurrence of human radiation in mid-Pleistocene in Europe. They were more advanced than *H. erectus* and were the intermediates between *H. erectus* and *H. sapiens*. The specimens discovered from Africa, Ethiopia and Zambia are believed to be the archaic *Homo sapiens* dating back 0.6–0.4 million years. The fossil jaws and the skull dating back to 0.78 million years from Spain are named as *Homo antecessors* and are deemed to be the common ancestors of Neandarthals and modern *H. sapiens*.

A parietal part of the skull from England (1935) was named as Swanscomb man and it resembled the modern man in size and curvature of bones. These are believed to be the intermediates between *Pithecanthropus* and modern man. *Homoneanderthalensis* (Neandarthal man) (Figure 17.4) were found in Germany (1848) as fragmented skull, ribs and limb bones with much more cranial capacity (1450 ccm) than that of the modern humans (1360 ccm).

Figuer 17.4 Neandarthal man

They had large eyes, broad nose and massive jaws and teeth. These are considered as the ancestors of *H. sapiens* as they were merely a coarsely built race of *H. sapiens*. But it is evidenced that Neandarthals seperated from modern humans and became established in Europe, Asia, Middle East, Britain and Spain before 0.5 million years so that they could be recognized as separate species. The DNA sequencing of an ancient human bone in 2008 from Southern Siberia indicates that the Neandarthals and modern men share a common ancestor about one million years ago. They were cultured and advanced in many ways. They made a variety of tools and weapons. They made clothes from animal skins, used fire, lived in caves or in shelters made of bones or skins. They seem to have disappeared about 30,000 years ago even though they coexisted with modern humans for several thousand years.

The fossils of modern *H. sapiens* were known from several parts in Africa and Israel. The early forms might have appeared in Europe during late Pleistocene about 4000–50000 years ago. They were called as Cromagnon man (Figure 17.5) who used advanced upper Palaeolithic tools. It is probable that the human evolution extends over a period of about 5 million years during which *Australopithecus africanus* evolved into the genus *Homo* that eventually gave rise to a variety of *H. sapiens* evolving into the modern humans.

Figure 17.5 Cromagnon man

MAJOR TRENDS IN HUMAN EVOLUTION

The hominid fossils throw light on several trends in cranial capacity, bipedalism, facial angle, reduced sexual dimorphism and so on during the process of human evolution.

a) ***Cranial capacity*** The first and early hominids with ape-like physique had moderate cranium which gradually increased in size. In the early Australopithicins, the cranial capacity was around 400 ccm as in the modern chimpanzees and in the later forms, it was 500 ccm. The early *Homo* species began with the cranial size of 510 ccm which increased well within the moderate range upto 1225 ccm in *H. erectus*. In Neandarthals, the cranial capacity was larger (1400 ccm) than that of the modern man (1350 ccm) due to larger body mass whereas the brain capacity of *H. sapiens* was greater.

According to Broadhurst et al. (1998), the nutritional resource in the African Rift Valley is the main creteria for the enlargement of the human brain. The great apes (chimpanzees or gorillas) are totally herbivores but the hominids might have included fish in their diet. As the lakes in the East African Rift Valley were alkaline and salty and supported tropical fish population at that time, the nutrient contents of the fish could be attributed for the greater development of brain and increased intelligence in *Homo* species. This is supported by a large collection of fish bones near a settlement of *Homo* dating back to 1.95 million years at the Rift Valley of Africa from where the humans are believed to have originated.

b) *Bipedalism* Bipedalism seems to have arisen in hominid line around 5–8 million years ago when man diverged from African apes. The conversion of forests into terrestrial wastelands in Africa near the end of Miocene period would have made a shift from tree-climbing quadrupedalism of the apes into the terrestrial bipedalism in humans. The fossil evidences suggest that the human evolution followed a 'locomotion first' pattern involving bipedalism. In apes, the foot is of grasping type with curved phalanges and opposable big toes. The pelvis is long and the backbone has a c-shaped curve. The skull is placed in front of the vertebral column. The occipital condyles and foramen magnum are placed behind the skull. In humans, the foot is a flat structure with a non-opposable big toe and the toes are provided with straight phalanges. The pelvis is short and the backbone has a s-shaped curve to have an erect or upright posture. The skull is placed on the top of the vertebral column and the occipital condyles and foramen magnum are placed beneath the skull.

c) *Skull morphology* In great apes and early hominids, the skull had distinct crests and ridges to provide anchorage for the neck muscles. The modern men have complete upright posture as the foramen magnum is centrally placed. But in early Australopithicins, the position of foramen magnum is intermediate between humans and apes so that their posture was not completely upright. In apes, the dental arcade is rectangular and right and left cheek teeth are arranged in two parallel lines. On the other hand, the dental arcade of *Homo sapiens* is in the form of a full parabola and is broader at the back than at the front.

d) *Facial angle* *Homo sapiens* exhibit a trend towards flatter facial angle with the resultant vertical face which is less projected with chin.

e) *Nature of rib case* The rib cases were funnel-shaped (narrow at the top and expanded towards the base) in Australopithicins whereas *H. erectus* and *H. sapiens* possess barrel-shaped rib cases.

f) *Lessening of sexual dimorphism* In Australopithicins, a marked degree of sexual dimorphism was evident but in the fossils of the early *H. Sapiens*, a lesser extent of sexual dimorphism could be traced.

g) ***Cultural growth*** The human cultural expansions include tool making, controlled use of fire, fabrication of shelters and clothings, appearance of art, development of co-operative hunting behaviour, domestication of plants and animals and so on.

In short, the process of human evolution involve the major events such as bipedalism, enlarged brain, use of stone tools and fire, wide geographical distribution, agriculture and beginning of global population increase. However, the causes for the above trends in human evolution are highly imaginary and speculative. The shortage of food, absence of rain, diminution of forests, problems in taking care of young ones and sexual activity, competition with other fauna, plenty of available food on the ground, etc., can be attributed for the appearance of terrestrial life from the arboreal habitat.

CULTURAL EVOLUTION OF MAN

The human evolution is strongly influenced by the biological and social processes as well as by culture. The biological evolution involves self-preservation, reproduction and greed which arose from millions and millions of years. When our primate ancestors approached closer to humans, less influence was exerted on our behaviour by the biological evolution but more cultural evolution was found to take over. As observed by Leaky (1978) this does not imply the end of the biological evolution but it simply altered the direction, i.e., the emerging human mind now evolved to fit its cultural niche to survive as a social creature.

Man has been a tribal animal since his first attainment of the erect posture more than four million years ago. For two million years, the early hominids were herd/tribal animals, especially herd herbivores. During the next two million years, they were tribal hunters/warriors. All human social drives like mother-love, compassion, co-operation, curiosity, inventiveness and competitiveness are instinctive and ancient as man attained all these long before he developed intellectuality.

The protohumans might most probably have had a relaxed life and eaten plant foods. But at the event of meat availability, they began gathering the food and bringing it back to a particular place. Now they distributed it among the other members who gathered around for a share. This would have been an adaptation for the improved survival ability. Initially they were scavengers as they were unable to succeed as hunters. At the event of continued cultural evolution, the limitations became overcome by co-operation resulting in great hunting ability. During food gathering, a division of labour arose among the members. The males went in search of meat and gathered the food supply whereas the females foraged for the plant foods. The task of hunting was difficult for the females, as the biological need required reproduction and responsibility of bearing and rearing the young ones. As a result, the females were biologically more valuable than the males who aided the females in reproducing. Thus the sharing of resources and division of labour had led to a divergence in biological and cultural evolution of the both sexes.

Man has been described as a 'tool-using animal'. The earliest known evidence for the manufacture and use of stone tools by humans (Stone Age) dates back to 2–5 million years in Africa. One of the earliest significant human activities was tool making with the aid of thinking power and manipulation of hands. This seems to be started two million years ago among the Australopithicines. At this stage, the food habit was also changed from vegetation to meat eating resulting in hunting practices. He engaged in developing tools and appliances because of his free hands and well developed brain. They made special weapons for the organized way of hunting. This resulted in social interactions and the need for communications followed by the development of languages.

The nature of tool development includes three major levels during the evolution of man. The Oedoman tool culture (Palaeolithic Age or Old Stone Age) associated with *H. habilis* (first tool makers) seems to be persisted in Africa for millions of years. The tools of this type are described as 'Cobble tools' which were of two main types namely flake tools (stones with one or more flakes) and core tools (stone tools without flakes). The more refined tools known as Acheulean tools (Mesolithic age) found in Europe and Asia are associated with *H. erectus*. These tools were advanced than the cobble tools and they had sharper and straight edges. During this period, the food habits were also changed in *H. erectus* who took meat on large scale regularly whereas the *Australopithecus* and *H. habilis* were mainly vegetarians. This is because *H.erectus* experienced fluctuations of temperature in earth's climate. The Ice age forced him to depend less on plants and more on animals. This change of diet gave man more time to spend in more different ways. He had an increased cranial capacity with greater thinking power and intelligence. The hides were used for clothing and shelter and the bones and antlers were used to make tools. *H.erectus* were the first hunters and used fire in a controlled manner. This could be a major step for social interaction in our ancestors. Thus they could have developed the ability to adapt for the changing conditions and improved methods for hunting and gathering food. The tool culture associated with Neandarthals and earlier *H. sapiens* is known as Monsterian tool culture. They used accurately shaped core stones with flakes modified for various functions. The modern *H. sapiens* used refined complex tools of various kinds made of wide range of materials. This is Neolithic culture or New Stone Age.

The archaic *H. sapiens* buried their dead bodies and used stone-tipped spears, fish hooks and so on. The modern humans used bows, arrows and other tools and weapons for hunting and fishing. They lived in well constructed huts and domesticated plants and animals. In short, they lived as hunter-gatherers in band societies. The Neolithic Revolution resulted in the advent of agriculture with permanent human settlements, domestication of animals and use of metal tools. This encouraged the trade amidst complex society with the resultant 'Human Era'. At present, man's efforts are mainly concerned with various cultural and technological advancements.

Thus man made chipped stone materials and weapons during Pleistocene times which was known as the 'Old Stone Age' (Palaeolithic Age). This is followed by the 'New Stone Age' (Neolithic Age) about 15000 years ago during which man was able to make and use polished stone materials in addition to the chipped stone tools and weapons. About 5000 years ago, he learned to use metals and this period was referred to as the 'Bronze Age' which was followed by the 'Iron Age' at the time of the great Old World Civilizations. At present, we are in the 'Steel Age' as well as in the 'Atomic Age'.

Culture is possible by the existence of mechanisms of learning and communication. In man, the major cause for cultural evolution seems to be the acquisition of intelligence. The behavioural decisions through the information learned by the members would have passed on to generations and established by nature. Thus the cultural evolution of man could not be a genetic one but it would be external favoured by genetic evolution after it had been established. Development of civilizations have influenced the human evolution and the ancestors of man have acquired a well developed brain which was an adaptive mechanism of power and efficiency. Thus man has got a mastery over the environments on the earth.

FUTURE EVOLUTION OF MAN

Human evolution includes two main phases namely a pre-human phase in Africa where the upright walking had evolved prior to two million years with many other characteristics still ape-like and a human phase with an increase in both brain size and behavioural complexicity and an expansion from Africa. In the case of man, the evolutionary process seems to be kept stand still so that it is very difficult to predict the future evolution of man with certainity. At the time when *Homo sapiens* evolved, several animal groups became extinct in the past due to many factors like interspecific competition, predators, drastic environmental changes and so on. Man underwent the most extreme changes in the body shape and continued to show genetically induced changes in physiology as well as in behaviour. Many typical animal behaviours such as mating, aggressiveness, etc., were inherited from his ancestors. He differs from other species in the use of advanced technology, social organisation and language.

Thus man has high adaptive value as he is an intelligent social animal with the ability to transform his environment and of high adaptibility to the changing environments. This has enabled him to become dominant with higher intelligence and efficient culture.

The most exciting aspect with regard to the future evolution of man is the possibility of the extinction of *H. sapiens* and the appearance of some other related species. This is highly impossible because the humans are free from unfavourable competitors, predatory attacks and drastic environmental changes. He has his own capacity to modify his environment and he can prevent death due to diseases and genetic disorders through his medical advancements. However, global wars and the appearance of the possible 'Super man' could be the probable chances for the extinction of human race. But these threats are much unlikely among the humans due to his cultural changes. It is ascertained that any adverse impact on humans could bring only redistribution and not the extinction. This is because man has split into few races and maintained unity in behavioural, physical and intellectual features.

Is there any possibility for man to speciate in future? The answer is that man has failed to speciate because of his great ecological diversity without chance to radiate into new habitats. Moreover, reproductive isolation is not prevalent among the humans to prevent gene flow between the races of man. Again the genetic innovations are only likely to become fixed in small isolated populations. Tattersall (2004) has opined that cross breeding could not be possible due to non-isolated populations to bring out potentially significant mutations in the gene pool as seen in the present day humans. Human populations are highly expanded and densely packed on earth as established communities. Therefore, any evolutionary possibilities is highly impossible in them. As noticed in the past, the organic evolution will be linked with the cultural evolution of man in future leading to drastic changes in earth's history. As such, the establishment of colonies on different habitable planets by man would possibly result in isolation for evolution of new human species in future. According to one view, Darwin's evolution is happening in humans on a very slow phase whereas other hypothesis believes that Darwin's evolution in humans is actually speeding up through sexual selections. Still others suggest that some major new isolating mechanisms would be needed for the appearance of a new human species.

Recently, it is presumed that the human brain is on the progressive growth in size so that the future man will be smarter and super intelligents. The fossil evidences indicate that there was no change in the human brain for 15000 years and then the *H. sapiens* were hunter-gatherers until about 10000 years ago with genes related to brain size, thus indicating that humans were still evolving. It is conceived that the process of enlargement of human cranium to accommodate larger brain has already started and the appearance of super intelligent man could be a matter of time. It does not mean that a new human species may replace the present humans but they would probably continue in another form. The brain size of future man could be increased three fold than the brain size of the present man. On the other hand, Chet Sherwood et al (2011) have hinted that the human brains are shrinking progressively with increasing age, a feature that is not seen in chimpanzees. This could be due to the loss of neurons leading to age-related neurodegradation at the cost of evolutionary process. According to Simon Laughlin et al (2011), the human brain may have reached its full capacity and the vast amount of extra energy cannot be provided by him to become more efficient so that he cannot become cleverer. However, others have viewed that the shrinkage of human brain would not bring any decline in man's intelligence but would make humans to be more intelligence.

In recent times, various opinions have been put forth with regard to the future evolution of man. Some are of the opinion that the lack of effective nervous system in industrial countries would lead to the deterioration of human species. But this could be counteracted by a number of factors such as immigration of people from less developed countries, the future use of genetic manipulation, the intentional breeding and so on. Anotherviewholdsthatthehumanracewillsplitintotwoseparatespecies namely an attractive intelligent and a dim-weighted ugly creatures. Oliver curry (2007) presumes that the human race will reach its physical peak by the year 3000 and then they will begin to regress. These humans are supposed to have a height of 6–7 feet and would live upto 120 years. In addition, the immune system could become weakened in them and they will begin to appear more child-like. It is also assumed that the future man would have big hairless heads without teeth but with larger brain, big eyes and thin weak body. They are supposed to possess only three fingers in the hands (thumb, index and middle fingers) due to non-functioning of organs on replacement of his work by machines.

In general, a species gradually evolves through Natural Selection into new species and this process would be very slow in a widespread and large species like humans. Human species can evolve without splitting into a new species via breeding or genetic manipulation. Then it is possible that *H. sapiens* could evolve through Natural

Selection or breeding or genetic manipulation. Moreover, evidences also indicate that gene mutations are also occuring today in man but as a very slow process. At present, man has enlarged cranium with the brain capable of designing more complex machines. It is also probable that man cannot survive without the help of machines in future. In short, it may mentioned that the scientific knowledge and advancements have bestowed man today with effective tools which seem to be the determining factors of future man.

QUESTIONS

1. Write a brief essay on the origin and evolution of man/*Homo sapiens*.
2. Give an account of fossil evidences in support of human evolution.
3. Enumerate the major trends in human evolution.
4. Write an account on the cultural evolution of man.
5. Discuss the future evolution of humans.
6. Write notes on the following :
   a) Anthropoidea
   b) *Dryopithecus*
   c) *Australopithecus*
   d) *Homo erectus*
   e) Neandarthal man
   f) Cromagnon man
   g) Stone age
   h) Tool culture of man

18

CONSEQUENCES OF CHORDATE EVOLUTION

Organic evolution is the process of continuous changes in the living world and it advocates that the present day highly organized complex living forms of the world have evolved from their ancestral stock that lived in the past through descent with modification. Those organisms which are surviving today would also undergo gradual and slow modifications leading to the origin of a variety of species. The evidences for the occurrence of organic evolution can be obtained from morphological studies of various organ systems of vertebrates. Though they are constructed on a basic plan, a few differences in some groups could be due to adaptive modifications to different modes of life (Homology). Certain organs which are similar in structure and function may be found in various groups with different pattern of development (Analogy). In some vertebrates, few structures are variously adapted for different modes of life but are constructed on the same fundamental plan derived from a single prototype. It means that the prototype model could have radiated in various directions to suite the environmental conditions (adaptive radiation or divergent evolution). Similar structural pattern with similar function found in distantly related groups of animals represents adaptive convergence (Convergent or parallel evolution). Thus, adaptations, adaptive radiation and convergent evolution can be regarded as the major consequences of chordate evolution.

ADAPTATIONS

The morphological or physiological modifications in an organism to live in a particular environment is referred to as adaptation. In order to survive in new circumstances, organisms have to change accordingly in the ever changing environment. Hence the environmental changes always result in the development of adaptive features which may be either morphological or physiological.

STRUCTURAL ADAPTATIONS

The changes in structure involving size, shape and external features of organisms or the anatomical modifications in the visceral organs mostly influenced by the physical environment are called structural adaptations. The following account provides the important and major structural adaptations seen in vertebrate groups.

Cursorial Adaptations

The modifications in the body of organisms for swift running constitute cursorial adaptations as seen in terrestrial animals. They depend on speed for getting food and for escaping from enemies. These animals possess several specialized features for fast running and are as follows:

1. The cursorial vertebrates possess stream-lined body with compressed thorax and flattened ribs which offer least resistance to air.
2. In the fore- and hind-limbs, the muscles are shortened and thickened and are found restricted in the proximal part of the limbs. The movement of the limbs are caused by the highly powerful and increasingly contractile muscles and spring ligaments found on the posterior region of the foot.
3. The limbs exhibit an overall increase in length due to the elongation of the carpals/tarsals and metacarpals/metatarsals of the fore-and hind-limbs respectively as well as due to an increase in length of digits. If the digit is elongated along the axis of the third toe as in horse, it is called mesaxonic pattern. If the lengthening is in between third and fourth toes, it is paraxonic pattern as in cattle, deer, sheep, etc.,

4. Some cursorial animals run on the digits of their feet to increase the speed (digitigrade). In these, the carpals and metacarpals of wrist and palm and the tarsals and metatarsals of ankle and sole are raised from the ground while walking or running as found in birds and most of the mammals. Some vertebrates like ungulates walk on the hoof which is the modified nail of the large and flattened bones of distal toes (unguligrade). These two types of postures seem to be derived from Plantigrade posture in which there are five toes in each foot. In digitigrade vertebrates, the number of toes are reduced to four whereas in unguligrade animals, the toes are reduced to four, three, two or one.
5. In cursorial animals, the ulna/fibula in fore- and hind-limbs as well as the clavicle in the pectoral girdles have become reduced.
6. The bones in the wrist and ankle are rigidly articulated and are incapable of bending so that the limb bones work as single unit for greater speed, e.g., lizards, flightless birds, marsupials, rabbits, horses, asses, zebras, dogs, cats, tigers, lions, etc.

Fossorial Adaptations

The burrowing mode of life is called as fossorial or subterranean or underground habitat. Those animals which live permanently in underground burrows with maximum structural modifications are known as total or true fossorial animals. On the other hand, some animals which dig the ground temporarily for getting food or retreat are called as partial fossorial animals. The true fossorial animals possess primitivc but marked organizational modifications because they experience static conditions with lesser light and enemies. The partial fossorial animals have simple modifications in the forelimbs. The following account provides the chief adaptations found in the true fossorial vertebrates.

1. The fossorial animals have cylindrical, compact, rigid,worm-like,fusiformorspindle-shapedbodywithpointedanteriorend. The head is very short and is conical and is often fused with the trunk. In these vertebrates, the neck and tail are absent

or reduced or vestigial. The eyes are very small, reduced and functionless. The pinnae or the external ears are absent or very much reduced but the ears are very sensitive to the sound waves which pass through the ground. In order to compensate the sight and hearing, the tactile organs are well developed in these animals.

2. The limbs are either absent or highly modified for digging and burrowing. The snout, forelimbs and teeth are the organs of digging. In most cases, the snout is often tubular and in some, the incisors and canines are modified for burrowing. The forelimbs are short, compact, strong and powerful. The fore arms are completely embedded in the body with broad and spade-like outwardly directed palm. All the digits of forelimbs possess powerful claws. The limb muscles are well developed as the forelimbs are useful for digging into the ground. The hind limbs help the animal to push into the tunnel.
3. The skull is conical, compact and is devoid of zygomatic arch with obliterated sutures between the skull bones. The jaw bones, dentition and auditory region are reduced. The burrowing animals mostly feed on insects and termites so that they have poorly developed jaw muscles.
4. The cervical, sacral and lumbar vertebrae are fused so that the vertebral column is rigid and firm, offering strength in burrowing.
5. The scapula and sternum of the shoulder girdle are elongated and strong and shift the forelimbs forward. The clavicles are thick and articulate with the scapula and humurus, thus strengthening the shoulders. A large T-shaped episternum and coracoids in the pectoral girdle provide additional strength to the shoulders.
6. The forelimb bones are short and strong in order to withstand the stress during burrowing. The humerus and ulna are short and stout with prominent muscle crests for the attachment of muscles. The tibia and fibula of the hind limbs are partially fused.

Eg : Burrowing fishes like *Amphipnous*, Caecilians, burrowing reptiles like Anguisnakes,Echidna, moles, armadillos, hedgehogs, squirrels, etc.

Arboreal or Scansorial Adaptations

Those animals which live on trees or climb in vertical surfaces are called as arboreal or scansorial forms. They include three groups namely 1) wall and rock climbers which are not actually arboreal forms but can climb on vertical surfaces 2) terrestrio - arboreal animals which can climb on the trees but are mostly found on earth and 3) true arboreal animals which mostly live on trees with frequent descent to the ground. The true arboreal forms may either be branch runners which are capable of running on the branches (squirrels, lemurs, chameleon, etc.) or suspended forms which hang head down (bats, sloths, etc.,) or swingers which can swing by the forelimbs (primates, apes, monkeys, etc.,). These arboreal animals have the following highly specialized structural features :

1. In general, the arboreal animals are smaller and slender. In some, the digits possess long and curved claws (bats, squirrels, lizards, etc.) or they may be long, curved and hook-like (sloths). Many have adhesive provisions like adhesive or suctorial discs in the digits (tree frogs), adhesive pads on the sole of the feet or dilated finger tips (tree-porcupine). In these animals, the tail is long and prehensile (chameleon, primates). In addition, some accessory organs such as parachute mechanisms (drago) or spines and tubercles (lemurs) are present in the fore arm. Some have grasping limbs with opposable digits (prehensilelimbs as in birds and arboreal reptiles and mammals). Syndactyli is another important feature in arboreal animals, i.e., the digits are arranged in two opposable bundles in both limbs facilitating to hold the branches.
2. The chest, ribs and limb girdles are strengthened with sub-circular thorax. The number of ribs are increased and the lumbar vertebrae are elongated and increased in number.

3. The pectorial girdle is heavily built and strong with prominent clavicle and scapula to support the weight of the body in tree branches. The ilium of the pelvic girdle is broadened.
4. In swingers, the humerus in the forelimbs is usually elongated whereas the digits are reduced (two or three digits only) in some cases.

   Eg: Almost every vertebrate group possesses arboreal representatives. *Anabas* and *Periophthalmus* (fishes), *Hyla* and *Rhacophorus* (Amphibia), Chameleon and tree snakes (Reptilia), sloths, tree Kangaroos, Civet cat, flying squirrels, flying lemurs (Mammalia).

Flying or Volant or Aerial Adaptations

The flying animals or volant forms are adapted for aerial mode of life and are represented in almost all vertebrate groups. Among them, some simply glide in the air for a very short period covering little distances (passive flight) and others are capable of sustained flight for longer durations (active flight).

Gliding adaptations The passive flight or gliding is found to be evolved independently in all vertebrate groups. There are no much specialized adaptations for passive flight in them. As an instance, the flying fishes have enlarged pectoral and ventral fins which are used as sail during gliding in the air. The flying frogs have webbed digits in the arms and are used as parachute. The flying lizard *Draco* possesses patagia in the trunk region between the fore-and hind-limbs with elongated ribs for gliding. All the flying mammals have parachute mechanism in the form of thin stretchable membranes between the fore-and hind - limbs. Thus the passive flight among the flying vertebrate representatives provides an example for analogy.

Flight or aerial adaptations The true flight involves sustaining of the animal in air for considerable distances with the help of organs of flight namely the wings. For this, both external and internal organization of these vertebrates have been modified markedly. However, the mode of development and structure of wings vary in different vertebrates suggesting that the flight have evolved independently in different groups. The general flight adaptations seen in pterodactyles, birds and bats include the presence of wings, stream-lined body, light weight due to pneumaticity, compact and rigid skeleton, strong flight muscles, high metabolic rate and so on.

Flight adaptation in birds Among the vertebrates, the birds are the aerial animals adapted both morphologically and anatomically for aerial mode of life. The following are the important flight modifications among the flying birds.

1. The body is spindle or boat shaped offering least resistance for easy propulsion through air. The light organs are positioned on the upper side of the body and the heavier ones on the lower side with wings attached high upon the thorax. The long mobile neck maintains low centre of gravity.
2. The forelimbs are modified as wings and the hind limbs are adapted to support the body on land. The unfolding and the upward and forward movement of wings result in upward propulsion of the body.
3. Presence of feathers is a unique feature in birds. The large quill feathers on the wings and tail are used for flight. The tail is very short but possesses a tuft of long tail feathers (rectrices) which act as rudder in assisting steering, lifting and counterbalancing. The entire body is covered by contour and filoplumes which prevent surface loss of heat due to radiation as they are non-conductors of heat (insulating structures).
4. The endoskeleton of birds is light but the strongest. The skeleton is rigid and compact due to the fusion of bones and is pneumatic because of the presence of air spaces. Therefore, the weight and specific gravity of the body are reduced.
5. The skull bones are light and are reduced in number with obliterated sutures. It has a single occipital condyle so that the head can be moved freely from side to side.
6. The thoracic vertebrae are fused so that the dorsal surface of the body is rigid to prevent dislocation and serves as a firm fulcrum aiding in wing strokes in air. The cervical vertebrae

have heterocoelous centra and they are highly mobile. The last thoracic, all the lumbar, sacral and few caudal vertebrae are fused to form synsacrum forming a plate-like structure which support the entire weight of the body. In addition, the posterior caudal vertebrae are fused to form pygostyle which not only provides attachment to the tail feathers but also acts as a rudder.

7. The pectoral girdle is rigid and the coracoids are heavy, long and stout. The flat scapula makes an angle with coracoid so that it resists inward pressure of the down strokes of the wings. The clavicles of both sides are fused to form v-shaped furcula which keeps the wings well apart as a spring. The sternum in the pectoral girdle is large with well developed keel for the attachment of flight muscles.
8. In the fore limb skeleton, the hand has three digits which are more or less fused and reduced. The pelvic girdle and the hind limb skeleton are modified to support the weight of the body.
9. The back muscles are reduced but the pectoral or flight muscles are well developed and attached to the keel of the sternum and to the wings. The largest muscle namely Pectoralis major causes the downward stroke of the wing. The thinner and longer Pectoralis minor is useful for the upward stroke of the wing. The small additional muscle, the Coracobrachialis also helps to pull the wings downward.
10. In birds, the internal organs are overlapped and closely packed. The digestive organs are capable of quick digestion in order to meet out the high rate of metabolism. The rectum is very much reduced and the faecal matter is not stored.
11. The lungs which are attached to the ribs are solid and spongy. They have a number of thin walled air sacs some of which communicate into the air cavities of the bones. The air kept inside the lungs and air sacs reduce the specific gravity of the body. The excess air inside the respiratory system facilitates complete aeration of the blood. The air sacs help in the internal respiration, which aids in the regulation of body temperature and offers buoyancy to certain limit.

The perfect oxygenation of blood imparts high temperature to the body so that the birds can fly at high attitudes and can be active in all seasons.

12. The heart in birds is four chambered, large and efficient for the separation of oxygenated and deoxygenated blood. It provides a large amount of oxygen to the tissues for rapid metabolism.
13. The kidney tubules and cloaca are efficient for the absorption of water. The urinary bladder is absent and the urine is immediately excreted, thus lessening the body weight by the excretory materials.
14. The brain is well developed in order to maintain balance and muscular coordination. The senses of touch and smell are poorly developed but the eyes and optic lobes are remarkably developed. The eyes possess comb-like pecten which enable the birds to see even very small animals from great heights. Moreover, the eyes are protected by the sclerotic bony plates which also play a role in adjusting the convexity of the eye balls.
15. Presence of a single ovary and oviduct also reduces the weight of the body.

Flight adaptations in bats Bats are the only mammals which fly in air with the aid of wings and they possess the following volant adaptations.

1. In bats, the wings are in the form of stretched extensions of skin supported by the forelimb bones between the fore - and hind-limbs (Patagia).
2. The forelimb bones are greatly elongated with long, thin and flexible matacarpals. The carpus freely allows flexion, extension and spreading of the digits.
3. The hind limbs with clawed digits are rotated outwards and the knee is directed backwards.
4. The sternum of the pectoral girdle has a well developed keel to give attachment to the well developed flight muscles as in birds. The stout clavicles are fused with sternum and scapula.

5. The energy required for the flight is furnished by the acquisition of a very high rate of metabolism.
6. Th bats are capable of echo-location system which is a special sensory system to locate the objects while flying even in dark. During flight, the bats emit ultrasonic sound waves with high frequency through which the animal can sense the objects through the pattern of the echoes like radar.

Aquatic adaptations The aquatic animals inhabit the water bodies so that their body organization is adapted to survive in aquatic medium. Those vertebrates which live permanently in water, including their ancestors are called primary aquatic animals and the animals which are secondarily adapted for aquatic mode of life are called secondary aquatic animals. The primary aquatic animals are the fishes which have evolved from aquatic ancestors and live permanently in water. They possess the following modifications in their body structure.

1. The body is boat or spindle-shaped or stream-lined which gives little resistance while moving in water.
2. The fins (paired, median and caudal) help to swim in water and also act as balancers and steering structures.
3. The body is covered by scales and the skin is provided with mucous glands.
4. The body muscles are arranged as bundles which enable lateral undulating movement of the body.
5. The gills are the respiratory organs for exchange of gases in water.
6. The air bladder or swim bladder filled with air in majority of teleosts acts as hydrostatic organ and helps for the upward and downward movement in great depths by controlling the inside air.
7. The specialized sensory system namely lateral-line system containing neuromast organs are the rheoreceptors used for the echo-location of objects inside water.

The secondary aquatic animals have evolved from terrestrial ancestors and they have returned to waters by developing aquatic mode of life due to unfavourable environmental conditions and heavy competition for food and shelter. As a result, they have acquired a number of changes in the body structure to lead successful life in water. While most of them exhibit amphibious life, some live permanently in aquatic medium. All these are lung breathers and possess the following secondary aquatic adaptations.

1. They have stream-lined body with short immobile neck and enlarged tail for rapid swimming in water.
2. The skin is smooth and naked without exoskeletal structures like armours, hairs, sweat and sebaceous glands, etc. To compensate this, some have developed a thick layer of fat beneath the skin (blubber) which helps to maintain constant body temperature, to avoid loss of body heat to reduce the specific gravity of the body and to act as a reservoir of food.
3. The external ears (pinnae) are absent (mammals) like seals, walruses and whales or reduced (amphibians). The external nostrils have moved to the tip of the head and are capable of being closed by valves. The eyes have flat cornea and rounded lens without nictitating membrane.
4. In amphibians and aquatic birds, the limbs are more or less webbed and the forelimbs are larger than the hind limbs for easy propulsion in water. In mammals, the hind limbs are reduced or lost whereas the forelimbs are modified as paddles or flipper. In some mammals, the tail is provided with caudal fin which helps to propel the body forward.
5. The bones are generally light and spongy. The skull is short and dorsoventrally flattened to accommodate the short and wide brain. In some, the skull is prolonged anteriorly to form elongated snout or rostrum for the easy capture of the prey. The zygomatic arch is reduced in some individuals. The teeth are simplified and mastigation power is lost in aquatic mammals. The jaws are usually adapted to catch hold of the prey and to prevent their escape.
6. In these vertebrates, the cervical vertebrae are fused into a bony mass without atlas and axis. The vertebral column is simple with reduced centra and zygopophyses in the vertebrae. But the vertebrae possess long processes in the tail region for the attachment of well developed muscles. The sacrum is reduced as there is no need for the support of hind limbs.

The single headed ribs are arched dorsally so that the body becomes cylindrical with increased thoracic cavity.

7. The short and compact hind limbs are articulated firmly without regional free movements of bones. To enlarge the surface area of the paddles, the limb bones exhibit hyperdactyli (extra digits) and hyperphalangi (extra phalanges). In aquatic mammals, the dorsal and caudal fins are fleshy and are not supported by fin rays.
8. The teeth are simplified and mastigation power is lost in aquatic mammals.

Example:Amphibians, turtles, crocodiles, geese, ducks, hippopotamus, etc. are amphibious and whales, porpoises, sea lions, dolphins, etc., are completely aquatic.

The changes enabling to inhabit new adaptive zones by a group of organisms is known as Preadaptations. In these animals, the modifications may either be beneficial or harmful in a particular environment in which they are already inhabiting. If the change is beneficial, then the animals could have been preadapted to the new environment. For example, the development of lobed fins, rigid back bones, lungs with internal nostrils, etc., are the preadaptations found in the crossopterygian fishes to inhabit the terrestrial environment. Therefore, this type of adaptation is also called as Prospective adaptation. In the new adaptive zone, these live successfully with the appearance of further adaptive characteristics.

In contrary, the development of adaptive features in animals to a particular environment in which they are already living is called postadaptations. These appear at the time of adaptive radiation and make the organisms to be fit for that environmental conditions. On the other hand, the modifications of animals in relation to the biotic environment (living organisms) are called coadaptations. Here, the organisms evolve to adjust each other as in the cases of commensalism, mutualism, parasitism and predator–prey relationships. In short, the organisms explore preadaptations which provide new and unoccupied areas for better opportunity and survival whereas the postadaptations make the organisms to be perfectly fit to a given environment. Thus adaptations, in general, permit the organisms to survive in the struggle for existence.

ADAPTIVE RADIATION (DIVERGENT EVOLUTION)

Evolution of organisms in several directions from a common ancestral type is known as adaptive radiation. In other words, the original ancestral stock enters into new adaptive zones. This type of evolution ends in the formation of new species resulting in micro and macro evolutions. Here, certain animals develop some new adaptations so that they move away from their ancestral adaptive zone, multiply and become diversified. During this process, the animals get more specialized features in the new habitat resulting in the formation of new species. A number of such radiations are evident during the evolutionary history of vertebrates by acquiring some major features such as origin of jaws and paired fins in gnathostomes, origin of pentadactyl limbs and lungs in amphibians, appearance of cleidoic eggs in reptiles, development of wings in birds and so on.

This evolutionary divergence can be well documented by the study of limb structure of vertebrates. The primitive insectivorous and short-legged ancestors which walked with the soles of the feet on the land are considered to be the ancestors to the present day mammals. The modern mammals diverged in five lines with their limb structures modified for five different habitats. Some evolved towards arboreal life (squirrels, primates); another group acquired the aerial mode of life (bat); some others developed cursorial adaptations (horse, deer, etc.); the fourth one led to fossorial life; still others became aquatic animals. In all these mammals, the limb structure is modified for a particular mode of locomotion. Moreover, these limb types can be easily derived from a common pentadactyl limb of terrestrial mammals.

Adaptive radiation can also be illustrated in Darwin's finches of Galapagos Islands. Generally, the ground finches have strong, stout and conical beaks of different sizes adapted to crush the seeds of varying sizes. The cactus ground finches possess curved beak and split tongue to collect the nectar from the cactus flower and to feed on the pulp. The vegetarian bee finches have short, thick and parrot-like beak to feed on leaves, buds and fruits. The insectivorous tree finches have the beak similar to the vegetarian tree finches but they feed on beetles.The woodpecker finches have stout, straight and long beak to feed on the insects found in the barks.

The evolution of teeth in mammals is another example of adaptive radiation. In mammals, the dentition is heterodont type (incisors, canines, premolars and molars) but the premolars and molars show structural variations for different modes of feeding. The premolars are low crowned with few simple cusps for crushing the prey in insectivorous mammals whereas they are low crowned with multiple complicated cusps in carnivorous forms. In herbivorous mammals, the incisors are sharp for seizing and cutting the vegetation and the premolars and molars are either short-crowned or long crowned. The teeth are simplified in fish-eating forms but they are slightly curved and homodont in toothed whales. The teeth are absent in sperm whales.

CONVERGENT EVOLUTION (PARALLEL EVOLUTION)

The phenomenon of development of similar characteristic features in unrelated or distantly related groups to inhabit similar environmental conditions is referred to as convergent evolution. The well known adaptive convergence is the acquisition of wings for flight in insects, flying reptiles, birds and flying mammals. The cartilagenous and bony fishes, extinct ichthyosaurs and aquatic mammals possess similar characteristics like stream-lined body, fins,modified limbs and so on for aquatic mode of life. The parallel evolution plays a role in the formation of analogous features in different animal groups.

QUESTIONS

1) What is adaptation? Brief different kinds of adaptations and their importance in evolution.
2) Give an account of cursorial/fossorial/ arboreal/ volant adaptations.
3) In what ways, birds are adapted to aerial mode of life?
4) Explain the evolutionary significance of adaptive radiation.
5) Write notes on :
   a) Adaptations
   b) Arboreal adaptations
   c) Volant adaptations
   d) Fossorial adaptations
   e) Aquatic adaptations
   f) Adaptive radiation
   g) Divergent/convergent evolution
   h) Pre/Post/Coadaptations

APPENDIX

A) LIVING FOSSILS

The term 'living fossil' was used by Darwin 150 years ago while discussing the genera *Ornithorhynchus* and *Lepidosiren*. The live animals that were believed to have been extinct and were known only by their fossil forms are considered as 'living fossils'. A living fossil can be defined as an animal that lived during ancient times and still survives today on earth. The living fossils were known once by their fossilized forms but were later found to be alive in remote parts of the world. Therefore, a living fossil is a species that closely resembles its fossil ancestors. As a living fossil is the surviving representative of an archaic group, it need not necessarily retain all the primitive features of the ancestral lineage and may possess one to many derived characteristics. Therefore it can be said that the living fossils are animals that have changed little over their life time as a species. Many fossil plants and animals having living representatives of present day are found in different layers of rocks all over the world.

The living fossils have remained essentially unchanged through a large portion of geological times and survive today with many primitive characters. They are mysterious creatures that exist today more or less unevolved from their ancestors hundreds of millions of years. Palaeontologists have found representatives of living fossils from every major vertebrate animal groups. While great majority of animals have undergone great changes over time, the living fossils are found to be well adapted to the environment without the need for the change. Two major reasons could be attributed for the existence of living fossils. In the first place, some living fossil species not only live in relatively isolated habitats with no apparent competitors but also at the same time in stable habitats due to the stabilizing selection so that there will be no pressure for them to change. Secondly, the living fossils would occur due to the absence of genetic variations in them.

Significance of Living Fossils

1. The extinction of some living fossils long ago and the discovery of their recent presence in living condition provides great curiosity.
2. The status of living fossils basically depends on the degree of similarity between the living and fossil species.
3. They are not rare and some of them provide a check on evolutionary speculations based on fossil record.
4. The living fossils are indicative of the environmental conditions of the geological past (palaeo environments of the geological record) and the occurrence of repetitive unchanged animals in the palaeo environments.
5. Some organisms are largely unchecked in an unchanging environment due to stabilizing selection so that many environments of today could not be greatly different from the environments of millions of years.

Creationists and Living Fossils

Evolution is a progressive change because of mutations and natural selection. But the living fossils show little or no change for over millions of years. Hence the discovery of living fossils possess a significant problem for the theory of evolution as they provide evidence against organic evolution. Evolutionists claim that all the living beings we see today emerged spontaneously by one species changing into another. However, Creationists argue that a large number of living animals species have remained unchanged for millions of years and their present anatomical structures are exactly the same as they were long ago as seen in the case of living fossils. But this is not true because all the animal species have evolved or descended from various 'ancestral forms' and a large number of 'intermediate forms' have emerged and lived over the long process of transitions, thus showing that the evolution is gradual and steady. The Creationists may be either unaware of the currently known fossils or unwilling to believe it. The discovery of a number of transitional fossils bridging the major vertebrate taxa evidences that the evolution had occurred and the organisms are related to each other.

Examples of living fossils Since 1822, thousands of previously unknown animals have been found and many of them are known to be the living fossils. Some of them have been recognized from fossils before their living representatives were discovered. The following table gives the most important examples of living fossils which have not evolved since the period of their fossil ancestors.

Table 19.1 Examples of living fossils in each major phylum

Taxa	Examples
Fishes	Lamprey, Hag fish, *Acipencer*, Coelacanth (*Latimeria*), Lung Fish(*Neoceratodus*), Frilled shark, Bow fish (*Amia*), Mackerel, etc.
Amphibians	Purple frog, toads and Salamanders
Reptiles	Pig-nosed turtle, Alligators, Crocodiles and Tuatara (*Sphenodon*)
Birds	Mouse birds, magpie Goose, New Zealand Wren, etc.
Mammals	*Platypus*, Opossums, Tarsiers, Bats, Squirrels, Elephant Shrews, SumatranRhinoceros, Tapirs, Koala bear, Giant Panda, Mountain Gorilla,Pygmy Chimpanzee, Okapi (pre historic giraffes), etc.

QUESTIONS

1. What are living fossils? Explain them with suitable examples.
2. Discuss the role of living fossils in the process of evolution.
3. Write notes on:

   a) Living fossils b) Creationists and living fossils

B) CONNECTING LINKS (MISSING LINKS)

The earth came to existence between four or five billion years ago and the life evolved on the earth about 3.5 million years ago. Since then, approximately 15 million different species of organisms have evolved. The origin of various species on earth has been a gradual and extensively slow process requiring hundreds or even thousands of years. This process of slow and gradual change in animal species is called organic evolution, which emphasizes that all living things on earth are the results of descent with modifications from a common ancestor. The fossil remains are the proofs for this evolutionary process of the past and the fossil record contains a register of all the billion years of life on earth. If the living organisms were evolved from common ancestors in the past, then several series of links connecting different groups of organisms must be present. It is true that there are a large number of missing links throughout the animal kingdom.

The plants and animals with intermediate conditions and characteristics linking two closely related groups are called as connecting links. In other words, the living organisms having intermediate characters between two groups are considered as connecting links. These animals establish continuity in the series of animals proving that one group has evolved from the other. They are the direct evidences for the evolution showing that the higher and complex forms originated from simple organisms. For example, *Protopterus* is a connecting link between bony fishes and amphibians, *Seymouria* between amphibians and reptiles, *Archaeopteryx* between reptiles and birds, Monotremes between reptiles and mammals and so on. Those extinct organisms which had the characteristics of two different groups of animals, confirming the path of evolution are referred to as missing links or transitional fossils. These animals undoubtedly indicate that the higher groups of animals have evolved from the lower ones and have passed through some transitional stages in order to reach their present status (Macro evolution).

However, Darwin was of the opinion that the fossil record contained not much evidence for the occurrence of connecting links. According to Creationists, ambient fossil proofs do not contain transitional forms in any of the species. If at all they appear, they argued that they are not the transitional between the members of the two modern phyla but they are transitional forms between the species of two primitive groups.

Thus this view actually strengthens the creation concept. But the fossil record during Darwin's period was poorly known so that it is reasonable for the lack of transitional fossils bridging the major vertebrate taxa at that time. In addition, it is highly unlikely for any organism to get fossilized and so it is unreasonable to expect an absolutely complete fossil record. Therefore, the animals that stand between two groups separating them from each other are the proofs for organic evolution. Because, when a species evolves into something else, intermediate forms of the two must also exist. In brief, it can be ascertained that the fossil record is full of links between earlier and later species so that Creationism is not a theory but just a belief. On the other hand, the theory of evolution is a scientific one that explains the fact of evolution.

Examples of connecting links The following table briefs the major transitional fossils which form a link between the primitive and higher groups of vertebrates.

Table 19.2 Transitional fossils that form a link between the primitive and major higher groups.

From	To	Transitional fossils
Primitive jawless fish	Sharks, skates and rays	*Pikaia, Cladoselache, Hypodus, Heterodontus, Chlamydoselache.*
Primitive bony fish	Holostean fish	*Cheirolepis, Polypterus, Sturgeons, Semionotus.*
Holostean fish	Advanced Teleost fish	*Leptolepis, eel, Diplomystus.*
Primitive bony fish	Amphibians	*Cheirolepis, Osteolepis, Icthyostega, Icthyostegopsis, Pholidogaser, Eucritta.*
Amphibians	Reptiles	*Seymouria, Limnoscelis, Hylonomous.*
Reptiles	Birds	*Liboasaurus, Protoavis, Archaeopterys, Hesperornis, Ichthyoinis.*

Table 19.2 *(Continued)*

From	To	Transitional fossils
Reptiles	Mammals	*Lystrosaurus, Megazostrodon, Pelycosaur, Dimetrodon, Cynognathus, Tritylodon, Diarthrognathus, Morganucodon, Amphitherium, Monotremes.*
Early eutherian mammals	Primates	*Notharctus, Parapithecus, Propliopithecus, Aegyptopithecus, Limnopithecus, Dryopithecus, Ramapithecus, Australopithecus, Homo habilis, Homo erectus, Homo sapiens, Neandarthalensis.*

Questions

1. Give an account on the significance of "Connecting links".
2. Write notes on: a) Connecting links/Missing links b) Transitional fossils

C) DISCONTINUOUS DISTRIBUTION

Living organisms are seen wherever it is possible for life to exist but they do not occupy all the possible areas. They are restricted to a particular area known as range. This denotes the area of distribution of species in space. On the other hand, the distribution of animals in time specifies the evolutionary history of the animals in earth's history. This is called as geological distribution which can be studied only through the knowledge of fossils. Mostly, the individuals of a particular genera or species or closely related species are found all over the world. These are said to have continuous or cosmopolitan distributions as seen in crows, hawks, bats, rats, etc., which are uniformly distributed in their range. But in some cases, the individuals of the same species or related species inhabit only localized areas which are widely separated without any continuity. These organisms are designed to have discontinuous distribution. The cosmopolitan animals which have a wide range of uniform distribution are called as eurytopic. These possess generalized adaptations and can survive in all possible environmental conditions. The organisms which are found in restricted areas are known as stenotopic which are characterized by having specialized adaptations to a particular environmental condition.

The continuity of the distribution is broken by a number of barriers which are the influencing factors that prevent the normal distribution of animals on earth's crust. The following are the major barriers which restrain the animal migration :

Physical barriers These include topography, vegetation, large water bodies and land masses, distance, impurity, salinity changes and so on.

Climatic barriers These are the ecological barriers which include temperature, moisture and light.

Biological barriers These include food, enemies, parasites, sedentary habit and home range.

The continuity of distribution of a species is broken by uninhabited areas and is influenced by evolutionary factors such as time and place. For example, the higher placental mammals evolved only after the separation of Australia and Asia so that the animals could not reach

Australia. The organisms that are adapted to special conditions as in the case of amphibians cannot survive on migration (Innate factors). The environmental factors such as area, topography and climate also play role in discontinuous distribution of animals. There are four primary reasons for the distribution of animals in widely separated areas.

1. *Movement by sweepstake routes* (rare chance dispersal across a major barrier): There is a possibility for the animals to reach distant areas by sweepstake means such as rafting, floating and swimming, storms, etc.
2. *Submergence of intermediate areas* The land masses in the intermediate areas could have submerged breaking the continuously distributed earlier species into widely separated populations.
3. *Continental drifts* The continents might have been separated and carried to long distances by continental drift, thus isolating the animals from other areas.
4. *Extinction in intermediate areas* A widely distributed species could become extinct in the intermediate areas due to change in the climate to which the species is unable to adapt.

Discontinuous distribution is well exemplified by the representatives of almost all invertebrate and vertebrate groups. The following table gives an account of vertebrate animals which are found in various parts of the world, providing examples for discontinuous distribution.

Table 19.3 Representatives of vertebrate animals exhibiting discontinuous distribution.

Animals	Habitat	Country
Lung fishes (Dipnoi)		
Protopterus	Samps and marshes	Africa
Lepidosirien	Marshes	South America
Neoceratodus	Pools and rivers	Australia

Limbless amphibians (Apoda)	Subterranean	Africa, South Africa, East Indies and Tropical America

Table 19.3 *(Continued)*

Animals	Habitat	Country
Alligators	Rivers and lakes	America and China
Flightless birds (ratitae)		
Ostriches	Desert	Africa and Arabia
Rheas	Plains	Argentina and South America
Cassowaries	Jungles	New Guniea and Australia
Emus	Grassy plains and open forests	Australia and New Guinea
Kiwi	Damp forests	New Zealand
Monotremes	Streams and deep forests	Australia
Marsupials	Varied habitats	South and North America, Australia and New Zealand
Tapirs	Plains	Central and South America, Malaya and Java and Sumatra islands
Camels		
True camels	Desert	Asia, North Africa, Mongolia and China
Llamas	Desert	South America
Elephants		
Elephants maximus	Forests and plains	India, Burma, Thailand and Sri Lanka
Loxodonta (Llamas)	Forests	Africa
Antelopes	Forests	africa

From the foregoing account, it is evident that some of the present day animals that are closely related seems to be distributed in widely separated areas. These animals would have had a wide range of distribution and were evenly found throughout the world in the past. Some natural barriers such as changes in earth's crust or climate or appearance of a variety of competitive organisms could have prevented the migration of these animals to mix with the same species. Thus the animal groups that are not able to cross the barrier become separated into isolated groups. Alternately, another possibility is that the present day distant places with same animals groups might have connected with each other by some physical means like bridges. In course of time, these connecting bridges might have submerged, thus separating the joined areas into two isolated places. For example, the lung fishes, Ratitae birds and marsupials had a worldwide distribution to begin with. Later due to continental drift, they became isolated. In places where competitions were severe, they were eliminated but they are able to survive even today in other places.

QUESTIONS

1. Write a note on discontinuous distribution with ample evidences.
2. Discuss various factors that govern discontinuous distribution.
3. Write notes on:
   a) Discontinuous distribution
   b) Reasons for discontinuous distribution.

D) CLASSIFICATION OF CHORDATES

Phylum-Chordata

Subphyla 1. Hemichordata - Balanoglossus
2. Urochordata - Ascidians
3. Cephalochordata - Amphioxus
4. Vertebrata or Craniata - Vertebrates

Subphylum-Vertebrata

Group I - Agnatha - Jawless Ostracoderms, lampreys and hag fishes

★ Class - Monorhina - Cephalaspids and Anaspids
★ Class - Diplorhina - Pteraspids and Coelolepids
★ Class - Cyclostomata - Lampreys and hag fishes
★ Class - Acanthodii - Acanthodians (spiny fishes)

Group II -Gnathostomata - Jawed Vertebrates

Super class - Pisces

★ Class - Placodermi - Early jawed fishes mostly with heavy armour (Climatius.)

Class - Chondrichthyes - Cartilagenous fishes.

Subclass - Elasmobranchii - sharks, skates and rays

★ Order - Cladoselachii - Cladoselache
★ Order - Pleurocanthiodii-Pleuracanthus
★ Order - Selachii - Sharks, dog fishes, skates and rays.

Subclass - Holocephali - Chimaeras and bradyodonts

Class - Osteichthyes - Bony fishes

Subclass - Actinopterygii - Ray-finned fishes

Infraclass - Chondrostei - Primitive ray - finned fishes (Acipencer)

Infraclass - Holostei - Intermediate ray - finned fishes (Amia)

Infraclass - Polypteryni - Polypterus

Infraclass - Teleostei - Advanced living ray - finned fishes

Subclass - Sarcopterygii - Air breathing lobe finned and lung fishes

★ Order - Crossopterygii - Rhipidistians, Osteolepids, Coelacanths,Latimeria, etc., (Lobe - finned fishes)

★ Order - Dipnoi - Lung fishes.

Superclass - Tetrapoda - Land living tetrapods

Class - Amphibia - The lowest and earliest land living tetrapods

★ Subclass - Labyrinthodontia - Ichthyostegids, Anthracosaurs, Embolomeres, Seymouriamorphs, Didectomorphs, Temnospondyls, Rhachitomes, Stereospondyls and Plagiosaurs

★ Subclass - Lepospondyli - Palaeozoic amphibians like

Microsaurs

Subclass - Lissamphibia (modern amphibians)

Order - Urodela or Caudata - Salamanders and Newts

(tailed amphibians)

Order - Salienta or Anura - Frogs and toads

Order - Apoda or Gymnophiona - Ichthyophis (Caecilians)

Class - Reptilia - scaled reptiles with amniotic eggs

Subclass - Anapsida - Reptiles with solid skull roof

★ Order - Cotylosauria - Primitive stem reptiles

★ Order - Mesosauria - Primitive aquatic reptiles

Order - Chelonia - Turtles and tortoises

Subclass - Diapsida - Ruling reptiles with two temporal fossae

and arcades

Infraclass - Lepidosauria

★ Order - Eosuchia - Primitive lepidosaurs

★ Order - Rhyncocephalia - Sphenodon

★ Order - Squamata - lizards and snakes

Infraclass - Archosauria - Advanced diapsids, like thecodonts*,

Crocodiles, Saurischian and Ornithiscian

dinosaurs*, Pterosaurs*, etc.

★ Subclass - Parapsida - Mostly marine reptiles with a single

fossa above thepostorbital - squamosal bones.

Order - Protrosauria - Primitive land - living reptiles

Order - Ichthyosauria - Fish-like marine reptiles

Order - Placodontia - Mollusc - eating reptiles

Order - Sauropterygia - Marine reptiles like Nothosaurs and

Plesiosaurs

★ Subclass - Synapsida - Mammal-like reptiles with a single

fossae below the postorbital - squamosal bones

Order - Pelycosauria - Primitive mammal-like

Ophiacodonts, Sphenacodonts,and Edaphosaurs

Order - Therapsida - Advanced mammal -like reptiles like

Anomodonts, Dicynodonts, Theriodonts, Cynodonts,

Ictidosaurs, etc.,

Class - Aves - Birds (flightless and flying)

★ Subclass - Archaeornithes - Archaeopteryx

Subclass - Neornithes - True birds, all living and few fossil

forms.

★ Super order - Odontognathae - Hesperornis and Ichthyornis

Superorder - Palaeognathae (Ratitae) - Flightless

birds

Order - Struthioniformes - African Ostrich

Order - Rheiformes - American Ostrich

Order - Casuariformes - Cassowary and Emu

Order - Apterygiformes - Kiwi

★ Order - Dinornithidea - Dinornis

★ Order - Aepyornithidea - Aepyornis

★ Order - Tinamiformes - Tinamus

Superorder - Sphenesciformes - Penguins

Superorder - Neognathae - Carinate or modern flying birds

Order - Pelecaniformes - Pelicans

Order - Ciconiiformes - Herons and strokes

Order - Anseriformes - Ducks, swans, geese

Order - Falconiformes - Eagles, vultures, falcons, hawks.

Order - Galliformes - Turkeys, Pheasants, quails

Order - Gruiformes - Cranes, rails

Order - Charadriiformes - Gulls, shore birds

Order - Columbiformes - Pigeons, doves

Order - Psittaciformes - Parrots, macaws

Order - Cuculiformes - Cuckoos

Order- Strigiformes - Owls

Order - Apodiformes - Swifts, humming birds

Order - Coliiformes - Colies

Order - Trogoniformes - Trogons

Order - Coraciiformes - Hornbills, hoopes, king fishers

Order - Piciformes - Woodpeckers

Order - Passeriformes - Song birds, perching birds

Order - Gaviiformes - Loons.

Order - Podicipediformes - Grebes

Order - Procellariiformes - Petrels

Class - mammalia - Tetrapods with mammary glands and hair

★ Subclass - Eotheria - Very primitive mammals

Order - Docodonta - Docodonts

Order - Triconodonta - Triconodonts

Subclass - Prototheria or Monotremata - Egg- laying primitive mammals

★ Subclass - Allotheria - Early extinct mammals

Order - Multituberculata - Primitive mammals like Plagiaulax

Subclass - Theria - Primitive and all advanced mammals

★ Infraclass - Pantotheria - Primitive Jurassic mammals (Symmetrodonts)

Infraclass - Metatheria or Marsupialia - Pouched mammals like opossums, bandicoots, kangaroo.

Infraclass - Eutheria - Placental mammals

Order - Insectivora - Most primitive placentals like shrews, hedgehogs and moles.

Order - Dermoptera - The flying lemurs (Galeopithecus)

Order - Chiroptera - Bats

Suborder - Microchiroptera - Small insectivorous bats

Suborder - Macrochiroptera - Large fruit - eating bats

Order - Primates - Lemurs, monkeys, apes and man

Suborder - Plesiadapoidea - Primitive ancestral Primates

Suborder - Lemuroidea - Lemurs

Suborder - Tarsioidea - Tarsiers

Suborder - Platyrhina - New World Monkeys

Suborder - Catarrhina - Old World Monkeys, apes, men.

Order - Edentata - sloths and armadillios

Order - Pholidota - Pangolins (Scaly ant eaters)

Order - Lagomorpha - Hares and rabbits

Order - Rodentia - Rodents like squirrels, rats, mice, porcupines (gnawing animals)

Order - Cetacea - Whales, dolphins, porpoises

Order - Creodonta - Early and persisting Creodonts

Order - Carnivora - Modern carnivorous placentals like cats, dogs, wolves, jackals, foxes, tigers, lions, leopards, bears, civets, hyenas, etc.,

Suborder - Fissipedia - Land living carnivores

Suborder - Pinnipedia - Marine carnivores (seals, walruses, sea lions)

Order - Ungulate - All hoofed mammals (condylarths, notangulates, South American ungulates, Odd-toed Perissodactyls like horse, zebra,rhinoceros, tapirs, even-toed artiodactyl - like pig, goat, sheep,hippopotamus, deer, ox, camel, giraffe, antilope, etc., (ruminants)

Order - Proboscidea - Largest living elephants, mastodonts, stegodonts, mammoths,Sirenians, etc.

Note : * denotes extinct forms

Acanthodians Acanthodians are the earliest known small vertebrates with true jaws resembling thesharks (and so the name 'spiny sharks') in late Silurian and were dominant duringearly Devonian and became extinct at the end of Palaeozoic era.

Acipencer It is the surviving genus of Chondrostei which evolved during late Palaeozoic eraexhibiting primitive, specialized and degenerate characters representing theChondrostean level in the evolution of bony fishes.

Adaptation The morphological or physiological changes in an organism to live in an everchanging environment for the survival in new circumstances are called asadaptations.

Adaptive radiation Evolution of organisms in several directions from a common ancestral type is knownas adaptive radiation by which the original ancestral stock enters into new adaptive zones resulting in the formation of new species.

Aegyptopithecus It was an Oligocene ape-like primate which resembled Proplipithecus and isconsidered to be the ancestors of Dryopithecus.

Agnatha A group that includes jawless vertebrates without paired appendages and comprises extinct armoured Ostracoderms and jawless living vertebrates like lampreys and hag fishes.

Anthrapogenesis The origin and evolution of man as a distinct species namely Homo sapiens is knownas anthropogenesis which is characterized by a number of important behavioural, morphological, physiological and developmental changes from an ape-like ancestorabout 5 million years ago.

Anthrapoids These are advanced primates, more highly organized than the lemurs and comprisemonkeys, apes and men and they might have originated from lemuroid ancestors during the upper Eocene and subsequent times.

Archaeopteryx It was a medium sized first known Jurassic fossil bird discovered from Bavaria(Germany) in 1861 with the resemblance both to a bird and a reptile, thus forming a'Connecting link' or 'missing link' between the thecodont Euparkeria and the modern birds.

Archaeornithes It is one of the subclasses of the class Aves and includes primitive and extinct reptile-like Jurassic bird such as Archaeopteryx which forms a connecting link between reptiles and birds.

Archaezoic era It is the 'period of primal life' with a time span of about 1000 million years,characterized by having abundant graphic iron ore and lime stone deposits but without a complete fossil record of organisms.

Arthrodires These are most varied Placoderms, commonly called as 'armoured fishes' or joint -necked fishes of Devonian period with domination during late Devonian and extinction at the end of Devonian period.

Australopithecus It was considered as ape-man which had characters of ape and man, thus forming a connecting link between the two groups so that these were the earliest forms with man-like characteristics leading to human evolution.

Cambrian period It is the oldest period of Palaeozoic era about 500–600 million years ago lasting for 80 million years with the rocks containing a large number of marine invertebrates andso called as the 'age of invertebrates'.

Carboniferous period Mississippian and Pensylvanian periods of Palaeozoicerarecombinedtogetherasthe Carboniferous Period about 230–320 million years ago lasting for 50 million years and characterized by the appearance of amphibians, winged insects,stem reptiles, dominance of marine echinoderms and the spread of ancient sharks and amphibians.

Cenozoic era It includes Tertiary and Quartenary Periods which are the periods of modern lifelasting for about 75 million years with the transition from ' the age of reptiles' into'the age of mammals' and characterized by the extinction of dominant reptilian groups and the rise and radiation of mammals.

Cephalochordates These are small fish-like marine chordates with notochord extending from head to tail but without distinct head or jaws or scales or paired fins, etc., thus possessing several primitive as well as specialized characters.

GLOSSARY

Ceratopsians These were the horned dinosaurs which were last to evolve in late Cretaceous where the earlier forms were hornless and the later individuals acquired a trend towards increased size and development of horns on the skull.

Chimaeras These are the very ancient shark-like fishes belonging to the group Holocephali,evolved from typical sharks during Devonian period and is known as fossils fromearly Jurassic times with a very less number of individuals persisting today.

Chiroptera It includes the bats which would have evolved from the arboreal insectivores during late Mesozoic or Tertiary times and they are the only mammals that have mastered true flight comprising Megachiroptera (most primitive large fruit - eating bats) and Microchiroptera (advanced small insectivorous bats including vampires, the modernblood - sucking bats).

Chordates These are coelomate, bilaterally symmetrical and triploblastic animals with distinct head and tail possessing segmentation, notochord, gill slits, dorsal tubular nervecord, ventral heart, etc.

Cladistics The phylogenetic analysis involving the search for monophyletic groups by means ofcharacter analysis.

Cladoselache It was the first known shark in the upper Devonian resembling some living forms oftoday with a number of primitive characters, thus considered as the central stem forthe later sharks.

Climatius It is an early small acanthodian fish from late Silurian and early Devonian times representing a generalized good example for ancient gnathostomes with a fish-like body.

Condylarths These were the most primitive ungulates which appeared at the beginning of the Cenozoic era, giving rise to a number of early ungulates as well as, two dominant ungulate groups namely Perissodactyla and Artiodactyla and also Probiscidea.

Connective links The plants and animals that possess intermediate features and establish continunity in the series of animals evidencing evolution are called connecting links.

Creodonts These were the archaic carnivores in Eocene times that evolved from the ancestral Mesozoic placental stock and all became extinct at the end of Eocene period and were regarded as the ancestors to the modern Carnivores as well as for the Cetaceans as they gave rise to Fissipede and Pinnipede Carnivores.

Cretaceous period It is one of the periods of Mesozoic era and was prevalent around 63–135 million years ago lasting for 60 million years with extensive fossil record and is characterized by the presence of marine reptiles, radiation of modern bony fishes (one of the major events in vertebrate evolution), extinction of dominant reptiles and primitive birds and the appearance of modern birds and primitive mammals.

Crossopterygians These were lobe-finned fishes which were abundant during Devonian times,declined in Carboniferous and became extinct before the end of Palaeozoic era andgave rise to the primitive amphibians during Mesozoic era.

Cyclostomes These are highly modified and degenerate survivors of some of the ostracoderm vertebrates such as lampreys and hag fishes possessing primitive as well as manyspecialized characters.

Dermoptera It is one of the mammalian groups comprising the flying lemurs with a single genusnamely Galeopithecus in which the skin extends between the neck and tail in the form of wing-like expansions.

Devonian period A period of Palaeozoic era about 345–405 million years ago lasting for 45 million years and is considered as 'the age of fishes' as many piscean species appeared and dominated and is also characteristic for the appearance of the first amphibians.

Didelphis It is an American Opossum which is a Cretaceous survivor of modern times representing the ancestral stock so that it is considered as a 'living fossil'.

Dinosaurs These were terrible lizards including a great variety of diversified groups of small andgreat giants which lived 200 million years ago for more than one hundred millionyears during Mesozoic era as two dominant major groups namely Saurischia and Ornithischia which all became extinct at the end of Mesozoic era.

Discontinuous distribution The occurrence of the individuals of the same species or related species onlyin localized areas which are widely separated without any continuity is known asdiscontinuous distribution.

Dryopithecus It was a late Miocene and early Pliocene ape that originated from Propliopithecus with a transformation from four-footed condition to bipedalism and erect posture so that this group would have given rise to man before the specialization of teeth as well as the modern apes.

Duck-billed dinosaurs These were large sized aquatic dinosaurs in which the front part of the skull andthe lower jaw were broad and flat like the bill of a duck with 500 or more number of teeth on the jaws for crushing the food.

Eosuchians These were the earliest lepidosaurs representing the first diapsids which gave rise toother lepidosaurian reptiles including Rhynchocephalia and Squamata.

Euparkeria It was the basic pseudosuchian in lower Triassic times with the appearance of archosaurian features like delicate and hollow bones, partial bipedalism, large opening at the sides of the lower jaw, sharp thecodont teeth, etc., thus indicating the evolution towards archosaurians.

Eusthenopteron It is a Crossopterygian fish which would be the direct ancestor to the first land vertebrates at the end of Devonian period during the transition from the Devonian to Mississippian times.

Fossils Fossils are the 'written documents of evolution' and are the preserved remains of hard parts of animals and plants that have become extinct and preserved intactwithout petrification.

Geological time The entire life span of earth is known as the geological time during the course of which the organisms exhibit a sequence of evolutionary events in definite intervalsand different durations.

Geological time scale The arrangement of divisions such as eras, periods and epochs on the time scale in the order of their age is called 'Geological time scale' which has been constructed based onthe correlations of the deposits all over the world through the study of fossils.

Gnathostomes These are the jawed vertebrates with paired appendages and include fishes,amphibians, reptiles, birds and mammals.

Hemichordates These are small, soft-bodied and worm -like lowest chordates having three body regions namely proboscis, collar and trunk with a rudimentary stomocord and Tornaria larva in the life history.

Homo erectus It was a human fossil with many features resembling the modern man but advance over *homo habilis*, so that it could be the most primitive human fossil and the missing link between the apes and man.

Homo habilis It was man-like fossil found in association with Australopithecus having a large brain capacity (630–700ccm) and hands possessing manipulating ability to make tools during Pliocene period, indicating that the human evolution had proceeded.

Homo heidelbergensis This fossil was considered to be the contemporarians of *Pithecanthropus pekinesis* which were related to Homo erectus and was more advanced than *H.erectus* and intermediate between *H.erectus* and *H.sapiens*, thus indicating theoccurrence of human radiation in mid Pleistocene.

Homo neandarthalensis It was a human fossil with much more cranial capacity than that of modern humans and considered as the ancestors of *Homo sapiens* but could have separated from modern man and become established as separate species.

Homo sapiens These were the fossils of modern man known as Cromagnon man that appeared during late Plestocene time from Australopithecus africanus which eventually gave rise to a variety of H. sapiens evolving towards the modern humans.

Hyoid arch It is the second

visceral arch in vertebrates and consists of a dorsal hyomandibular bone and the ventral hyoid cornu of which the former helps to suspend the jaws with the cranium, so that it played a major role in the evolution of fishes and tetrapods.

Hypodont sharks These are the primitive earliest representatives of modern sharks representing intermediate forms in the evolution of cartilagenous fishes possessing both primitive as well as advanced features exhibiting climax in the late Palaeozoic period.

Ichthyosaurs These were the most highly specialized marine Parapsids that evolved from Cotylosaurian ancestors exhibiting a single pattern of adaptation in their history and were considered as big fishes with first swimming adaptations.

Ichthyostega A primitive amphibian found in the upper Devonian period well on the way to the full pledged amphibians representing as intermediates between the osteolepid fishes andearly amphibians.

Insectivora It is the group that includes the stem placentals from which the eutherian mammals would have evolved through a number of independent lines and all the present day placentals might have evolved as offshoots of ancestral insectivores during Eocene period.

Jamoytius A very primitive jawless vertebrate belonging to Anaspid ostracoderm with small elongated tubular body occupying a position close to the ancestors of the lampreysand their relatives.

Jurassic period This period is one of the periods of Mesozoic era and was the 'Age of Giants' datingback to 135–181 million years lasting for about 30 million years as dinosaurs dominated the late Jurassic and is characteristic of origin of first toothed birds and appearance of lizards, crocodiles and very ancient mammals.

KT event The Cretaceous-Tertiary boundary is known as KT event and there was a mass extinction of a number of vertebrate groups especially the dinosaurs about 65 millionyears ago across the KT interval.

Labyrinthodonts These are the earliest known limbed extinct amphibians belonging to Stegocephaliaand are believed to be evolved from the Crossopterygian fishes leading to the primitive reptiles.

Latimeria It is the only living representatives of the group Crossopterygii, known from time totime in Madagascar and Africa so that it is considered as ' living fossil'.

Lazarus taxon A species or groups of species that suddnely reappears either in the fossil record or innature.

Living fossils The live animals that were believed to have been extinct and were known only by their fossilized forms are called as living fossils, i.e., these are the animals that lived during ancient times and survives even today on earth.

Lung fishes These include air breathing dipnoans which would have evolved as a side branch ofthe bony fishes from the primitivecrossopterygiansduring middle Devonian, occurring in plenty during Palaeozoic times with a decline except a few surviving representatives namely Protopterus, Lepidosiren and Neoceratodus in very much restricted areas like Australia, South America and Africa respectively, thus illustrating discontinuous distribution.

Mandibular arch It is the first visceral arch in vertebrates consisting of a dorsal palatoquadrate and aventral Meckels Cartilage which give rise to the upper and lower jaws respectively.

Marsupials These are the pouched mammals found in Australia and South America (discontinuous distribution), intermediates between monotremes and placentals butare more advanced than the monotremes with the origin from the Pantotheres during early Cretaceous times and specialized with an integumentary brood pouch (marsupials) enclosing the teats through which the immature young ones are fed withother's milk.

Mesolithic Age It was the period during which more refined tools advanced than the cobble tools with sharper and straight edges called Acheulean tools were used by Homo erectus.

Mesozoic era It is the period of medieval or intermediate life consisting of Triassic, Jurassic andCretaceous periods around 135–205 million years ago lasting for 130 million years and is considered as 'the Golden Age of Reptiles' due to the dominance andextinction of ancient reptiles and also characterized by the origin of birds andmammals.

Monotremes These are the egg-laying prototherian mammals inhabiting Australia and NewGuinea evolved from the docodonts during Triassic, forming intermediates between reptiles and mammals as they possess both mammalian and reptilian characters and represent living fossils as they are very primitive living mammals.

Multituberculates These were highly specialized extinct triconodonts adapted for herbivorous mode offeeding during the upper Jurassic to Tertiary and were considered to be the ancestorsof herbivorous mammals.

Neolithic evolution It denotes the advent of agriculture with permanent human settlements,domestication of animals and use of metal tools and the consequent encouragement of trade amidst complex society with the resultant 'Human Era'.

New stone age It is otherwise known as Neolithic Age during which accurately shaped core stoneswith flakes modified for various functions (refined and complex tools of various kinds made of wide range of materials) were used by the Neandarthals and earlierHomo sapiens and this culture is known as Monsterian tool culture.

Old stone age It is otherwise known as Palaeolithic Age in which the nature of tool developed was the Oedoman tool culture associated with Homo habilis (first tool makers) who used'Cobble tools' (flake and core tools).

Ordovician period A period of Palaeozoic era which existed 425–600 years ago lasting for about 65million years characterized by the appearance of the first fishes especially fresh water fishes and the presence of a number of invertebrate fossils.

Oreopithecus It was a Pliocene ape which was similar to man in many features especially in dentalnature so that it is believed that man might have originated from Oreopithecus.

Ornithischians These were the dinosaurs with tetraradiate pelvis as in birds and so the name Ornithischia and were slow walking reptiles on their four legs.

Ostracoderms The best known ancient agnathans originated in Cambrian times from somesuccessful group of Palaeozoic unarmoured ancestors about 550 million years ago and some resembling the modern hag fishes and lampreys and others with bonyarmour plates for protection.

Palaeozoic era It was the time of ancient life about 505 million years ago lasting for nearly 300million years with six periods namely Cambrian, Ordovician, Silurian, Devonian,Mississippian, Pennsylvanian and Permian based on the order of age.

Pantotheria It was one of the earliest known extinct Triassic and Jurassic mammalian groupwhich gave rise to two important mammalian groups namely marsupials and placentals during early Cretaceous prior to their extinction at the end of Cretaceousperiod.

Parapithecus It was one of the first Old World Monkeys (anthrapoid apes) which appeared during Oligocene and formed a link between the tarsoids and Old World Monkeys.

Pelycosaurs These were the first synapsids that appeared from Ophiacodont stem in latePennsylvanian to Permian times and most common land vertebrates during Permian times and evolved into two groups namely sphenacodonts and Edaphosaurs.

Permian period It is the youngest period of Palaeozoic era that lasted for 25 million years around 230–280 million years ago and is characterized by the appearance of mammal-like reptiles andthe extinction of several lines of amphibians.

Phylogency It is the evolutionary tree that indicates closeness of relationship among organisms.

Placoderms Placoderms are the primitive gnathostomes but more specialized than theacanthodians evolved during middle Palaeozoic representing a hetrogenous group and became extinct at the end of Palaeozoic era around 360 million years ago.

Platetectonics These are the processes beneath earth's crust that produce new crust along mid-oceanic ridges and cause oceanic and continental plates to move.

Precambrain era The Archaeozoic and Proterozoic eras have been collectively called as Precambrian or Cryptozoic era as these two eras existed prior to the beginning of the fossil recordand most of the fossils were lost due to changes on earth's crust.

Primates These are the last and most highly evolved mammals such as lemurs, tarsiers, monkeys, apes and men arising from insectivore ancestors during Cretaceous timesalong various lines with their establishment in the beginning of Cenozoic era.

Proboscideans These include largest land-living herbivorous mammals namely elephants that possess a proboscis with external nostrils at the tip (elongation of nose and upper lip),very large head, pillar-like legs, large external ears, tusks formed by the elongation ofincisors and small eyes.

Prolacerta It is the genus of the early Triassic period evolved from eosuchian ancestors andcould be the very near ancestor of lizards as it had cheek similar to the ancestral lizards but the skeleton retaining the basic eosuchian characters.

Propliopithecus It was ape-like primate from Oligocene times and was considered to be the direct ancestor for modern Gibbons.

Proterozoic era It is the 'period of primitive life' with a time span of about 1500 million years duringwhich algae, marine protozoans, molluscs, annelids and other marine invertebrates were found as fossils.

Pterosaurians These were the archosaurians adapted for flight at the beginning of Jurassic times and continued through Cretaceous period at the end of which they became extinct.

Quartenary period It is described as the 'Age of man' as the origin of man was the important event in this period characterized by glaciations with the duration of one million years and includes two epochs namely Pleistocene (end of the Ice Age 0.975 years ago with thedecline of many mammals and the start of social life in man) and Recent epoch (post -glacial period, and 'Psychozric era' representing the intellectual evolution about 25000 years ago as human society established social life).

Ramapithecus These were the Contemporarians of Oreopithecus in Pliocene times and were the perfect intermediates between apes and humans, but were more closely related to Orangutans and were not directly connected to human evolution.

Ratitae These are large flightless birds belonging to the group Palaeognathae found restricted to Africa, South America, Australia, New Guinea, New Zealand and Madagascar, thus exhibiting discontinuous distribution and they would have probably evolved from flying ancestors.

Relic An organism that persists in one area long after its relatives have become extinct elsewhere.

Rhynchocephalia It is one of the four orders of modern reptiles evolved from the thecodont ancestors at the beginning of Triassic period and is now represented by the single genus Sphenodon thus forming a very much restricted group with a limited evolutionaryline from Triassic to till date.

Rodents These include the gnawing mammals which resemble the insectivores more closely than any other mammals and they might have separated from the placental stock before the appearance of the insectivores.

Sarcopterygii A subclass in Osteichthyes comprising dipnoans (lung fishes) and Crossopterygians(lobe finned fishes) which are air breathers and occupy an important position in the evolutionary history of vertebrates as they played a significant role in the progress beyond fishes.

Saurischians These were the dinosaurs with triradiate pelvis having teeth on the jaws with the hindlimbs used for swift running and the forelimbs for procuring food.

Seymouria It is a primitive reptile evolved from labyrinthodont amphibians during lower Permian time and was almost an intermediate form or connecting link between the amphibians and reptiles.

Silurian period A period of Palaeozoic era with a time span of about 35 million years around 405–425 million years ago during which the terrestrial plants, wingless insects, marine arachnids and fishes appeared.

Skull It is the brain case or cranium which may be either cartilagenous (chondrocranium asin cyclostomes and elasmobranchs) or bony (neurocranium as in vertebrates) that contains the sense organs (eyes, nose and ears) and splanchnocranium or visceral skeleton comprising the visceral arches.

Sphenodon It is otherwise known as Tuatara or Hatteria or New Zealand lizard which is the sole surviving rhynchocephalian evolved at the beginning of Triassic period possessing many primitive as well as lacertilian and crocodilian features and is now confined to New Zealand so that it is considered as the 'living fossil'.

Stegosaurus These were large, earliest known secondarily quadrupedal camptosaur Ornithischians commonly known as plated dinosaurs from early Jurassic times with peculiar plates on the back and spines on the tail.

Stem reptiles The first reptiles from the primitive amphibian stock during Late Pennsylvanian andearly Permian times belonging to the group Cotylosauria with adadptations to lead asuccessful life on land possessing especially the land eggs and scales to give rise to allhigher reptiles between the Permian and Cretaceous times.

Stone age It was the time by which the humans manufactured and used stone tools about two million years ago so that tool making was one of the earliest significant human activities with the aid of thinking power and manipulation of hands and it seems to be started among Australopithicines.

Swanscomb man This fossil resembled the modern man in size and curvature of bones so that it isbelieved to be the intermediate between Pithecanthropus and modern man.

Terrestrialization The adaptations in the body acquired by the first land vertebrates during the transitional period from the aquatic life to live on land in Devonian times is known asterrestrialization.

Tertiary period It is one of the periods of Cenozoic era which lasted for 74 million years characterized by the origin and evolution of mammals and comprises five epochs namely Palaeocene (abundance and radiation of placentals lasting for 17 million years),Eocene (specialization and diversification of placentals and extinction of primitive egg laying mammals lasting for 19 million years), Oligocene (appearance of higher placentals and anthropoid apesand extinction of primitive mammals, lasting for 11 million years), Miocene (peak mammalian evolution and appearance of first man-like apes, lasting for 16 million years and Pliocene (dominance of elephants, horses and camels as well as origin ofman from apes, lasting for 11 million years).

The golden age of reptiles The mesozoic era ia known as ' the golden age of reptiles' as it was dominated by the huge dinosaurs and other reptiles on land, the flying reptiles or pterosaurs in air and fish and lizard-like swimming reptiles such as Ichthyosaurs, Plesiosaurs and other marine reptiles in sea.

Theriodonts These were the Triassic mammal-like primitive synapsid reptiles which would have evolved from the pelycosaurs and bridged the gap between the reptiles and mammals as they had both reptilian and mammalian features and were the central stock for the mammals at the end of Triassic.

Tree shrews These were the first primates evolved from insectivores and were the ancestors for the higher primates such as monkeys, apes and men (Anthrapoidea).

Triassic period It is the earliest division of the Mesozoic era about 181–230 million years ago lasting for 40 million years constituting a period of transition between the life of Palaeozoicera and Mesozoic times in which the fishes, amphibians and reptiles progressedtowards the future evolution of vertebrates so that various groups from the Permian period lived side by side with numerous new reptiles with the evolution of dominantanimal groups of middle and later Mesozoic era.

Ungulates These are the hoofed mammals (Perissodactyla and Artiodactyla) that originated from insectivore ancestors during the transitional times between Cretaceous and Tertiary evolving along fourteen lines during Cenozoic era so that they are the last but a greatgroup of placentals.

Urochordates These are marine and mostly sedentary chordates having notochord in the tail with free swimming larva possessing many chordate features like notochord, dorsal tubular nerve cord, gill slits, etc.

Vertebrates These are the higher chordates having cartilagenous or bony skull, jaws, vertebrae(vertebral column), paired appendages, head with sense organs, nervous system with brain and spinal cord, the internal ear with three semicircular canals, circulatory system with muscular heart, blood cells and blood vessels, etc.

Youngina It was a lizard-like earliest lepidosaur (Eosuchian) in the upper Permian, with many Cotylosaurian features so that it would have given rise to the later diapsid reptiles including the modern lizards and snakes.

REFERENCES

Berril, N.J. (1955) *The origin of Vertebrates*, Oxford Uni.Press, London.

Brooks, C.E.P. (1970) *Climate through the Ages* (2nd ed), Dover Publications, Inc., New York.

Carroll, R.H. (1988). *Vertebrate Palaeontology and Evolution*, W.H. Freeman and Co, New York.

Carroll, R.H. (1977). *Patterns and Process of Vertebrate Evolution* (1st ed), Cambridge Uni.Press,New York.

Colbert , E.H. (1980). *Evolution of the Vertebrates* (3rd ed), John-Wiley and Sons, New York.

Lull, R.S. (1947). *Organic Evolution*, MacMillan Co., New York.

Moody, P.A. (1953). *Introduction to Evolution*, Harper and Brothers, New York.

Neilson, C. (2001). *Animal Evolution-Inter relationship of the living Phyla* (2nd ed), OxfordUni. Press Inc., New York.

Olson, E.C. (1971). *Vertebrate Palaeontology*, John Wiley and Sons, Inc., New York.

Romer, A.S. (1966). *Vertebrate Palaeontology* (3rd ed), Uni. of Chicago Press, Chicago.

Simpson, G.G. (1953). *The Major Features of Evolution*, Columbia Uni. Press, New York.

Smith, H.M. (1960). *Evolution of chordate structure—An Introduction*, International Thomson Publishing.

INDEX

P

R

S

T